ALGEBRAIC FOUNDATIONS OF MANY-VALUED REASONING

TRENDS IN LOGIC
Studia Logica Library

VOLUME 7

SCOPE OF THE SERIES

Trends in Logic is a bookseries covering essentially the same area as the journal *Studia Logica* – that is, contemporary formal logic and its applications and relations to other disciplines. These include artificial intelligence, informatics, cognitive science, philosophy of science, and the philosophy of language. However, this list is not exhaustive, moreover, the range of applications, comparisons and sources of inspiration is open and evolves over time.

The titles published in this series are listed at the end of this volume.

ROBERTO L.O. CIGNOLI
Department of Mathematics,
University of Buenos Aires, Argentina

ITALA M.L. D'OTTAVIANO
Department of Philosophy and The Centre for Logic,
Epistemology and the History of Science,
State University of Campinas, Brazil

and

DANIELE MUNDICI
Department of Computer Science,
University of Milan, Italy

ALGEBRAIC FOUNDATIONS OF MANY-VALUED REASONING

Springer-Science+Business Media, B.V.

A C.I.P. Catalogue record for this book is available from the Library of Congress.

ISBN 978-90-481-5336-7 ISBN 978-94-015-9480-6 (eBook)
DOI 10.1007/978-94-015-9480-6

Printed on acid-free paper

To the memory of
ROLANDO CHUAQUI
ENNIO DE GIORGI
ANTÓNIO MONTEIRO
great Scientists and Teachers

Contents

Introduction

The aim of this book is to give self-contained proofs of all basic results concerning the infinite-valued propositional calculus of Łukasiewicz and its algebras, Chang's MV-algebras. This book is for self-study: with the possible exception of Chapter 9 on advanced topics, the only prerequisite for the reader is some acquaintance with classical propositional logic, and elementary algebra and topology.

In this book it is not our aim to give an account of Łukasiewicz's motivations for adding new truth values: readers interested in this topic will find appropriate references in Chapter 10. Also, we shall not explain why Łukasiewicz infinite-valued propositional logic is a basic ingredient of any logical treatment of imprecise notions: Hájek's book in this series on Trends in Logic contains the most authoritative explanations. However, in order to show that MV-algebras stand to infinite-valued logic as boolean algebras stand to two-valued logic, we shall devote Chapter 5 to Ulam's game of Twenty Questions with lies/errors, as a natural context where infinite-valued propositions, connectives and inferences are used. While several other semantics for infinite-valued logic are known in the literature—notably Giles' game-theoretic semantics based on subjective probabilities—still the transition from two-valued to many-valued propositonal logic can hardly be modelled by anything simpler than the transformation of the familiar game of Twenty Questions into Ulam game with lies/errors.

This book is mainly addressed to computer scientists and mathematicians wishing to get acquainted with a compact body of beautiful results and methodologies—that have found applications in the treatment of uncertain information, (e.g., adaptive error-correcting codes) as well as in various mathematical areas, such as toric varieties, lattice-ordered groups and C^*-algebras. As the title indicates, the main emphasis is on algebraic methods. Thus, reversing the historical order, we shall make the reader familiar with MV-algebras before introducing Łukasiewicz's propositional calculus in Chapter 4. This will allow us to get neat and elementary proofs of several deep results, using much less symbolism and detail than in traditional syntax-oriented approaches. The definition-theorem-proof style adopted throughout this book will hopefully result in time saving for the reader who wishes to get the proofs of all main theorems on the infinite-valued calculus as quickly as possible, without embarking on a potentially unbounded search through a scattered literature on ordered groups, lattices, algebraic logic, polyhedra, geometry of numbers, model theory, linear inequalities, et cetera.

By definition, an MV-algebra A is a set equipped with an associative-commutative operation $\oplus$, with a neutral element 0, and with an operation $\neg$ such that $\neg\neg x = x$, $x \oplus \neg 0 = \neg 0$, and, characteristically,

$$\neg(\neg x \oplus y) \oplus y = \neg(\neg y \oplus x) \oplus x.$$

These six equations are intended to capture some properties of the real unit interval $[0, 1]$ equipped with negation $\neg x = 1 - x$ and truncated addition $x \oplus y = \min(1, x + y)$. For instance, once interpreted in $[0, 1]$, the left hand term in the last equation coincides with the maximum of x and y; thus the equation states that the max operation over $[0, 1]$ is commutative. The fundamental theorem on MV-algebras is Chang's completeness theorem, stating that every valid equation in $[0, 1]$ is automatically valid in all MV-algebras. A new proof of this theorem is given in Chapter 2. As a preliminary step, in Chapter 1 we prove Chang's subdirect representation theorem, stating that an equation is valid in every MV-algebra iff it is valid in every totally ordered MV-algebra.

As in the classical case, one may ask for an effective procedure to decide when an equation is valid. Rather than working in "MV-algebraic equational logic", it is more convenient to give the Łukasiewicz infinite-valued calculus the same role that the classical propositional calculus has for the boolean decision problem. Accordingly, one may write

$$\neg x \rightarrow y \text{ instead of } x \oplus y$$

and transform valid MV-equations into tautologies by writing $p \leftrightarrow q$ instead of $p = q$. The main theorem proved in Chapter 4 then states that the rule of modus ponens is sufficient to obtain all tautologies in the infinite-valued calculus of Łukasiewicz starting from four basic tautologies (originally given by Łukasiewicz) corresponding to the above defining equations for MV-algebras. These equations are thus "complete", in the sense that every equation that is valid for $[0, 1]$ is obtainable from them by substituting equals for equals. As another corollary, in Chapter 5 we shall show that tautologies in the infinite-valued calculus of Łukasiewicz coincide with those formulas that are true in every Ulam game, independently of the number of errors/lies.

Having thus acquired a unified view of valid MV-equations and infinite-valued tautologies, we can handle logical notions using standard algebraic methods.

Thus, e.g., from the logic-algorithmic viewpoint, free MV-algebras over n generators consist of all equivalence classes f of formulas in n variables. On the other hand, from the algebraic-geometric viewpoint, free MV-algebras consist of all continuous $[0, 1]$-valued piecewise linear functions f with integer coefficients defined over the cube $[0, 1]^n$. (This is McNaughton's representation theorem. See Chapter 3 for the case of one variable, and Chapter 9 for the general case). The multiple nature of f is useful in the study of normal form reductions, and it yields a concrete visualization of the notion of consequence and its fine structure. Further, one can strengthen the completeness theorem and give an algorithm to decide whether a formula is a tautology—having no greater complexity than its analog for boolean tautologies (see Chapter 9).

Generalizing the relationship between the interval $[0, 1]$ and the naturally ordered additive group of real numbers, every MV-algebra A can be realized as the unit interval $A = [0, u] = \Gamma(G, u)$ of a *unique* abelian lattice-ordered group G with a strong unit u equipped with negation $u - x$ and truncated addition $x \oplus y = (x + y) \wedge u$. Specifically, as proved in Chapter 7, Γ is a categorical equivalence between abelian lattice-ordered groups with strong unit, and MV-algebras. Among the many important consequences of this equivalence, one can unambiguously say, e.g., that elements $a_1, \ldots, a_n \in A$ "sum up to one", or that they are " linearly independent". These two conditions allow one to give an MV-algebraic definition of "partition of unit", thus generalizing the basic notion of boolean partition.

Since abelian lattice-ordered groups are so well established—their roots going back to the time-honored theory of magnitudes—one might wonder, why MV-algebras should be given special attention. One main reason is that, while a strong unit u in a group G of magnitudes is no less important than the zero element, the property of u being a strong unit in G is formalized by

$$\text{for all } x \in G \text{ there is } n = 0, 1, 2, \ldots \text{ such that } x \leq nu.$$

This archimedean-like property is not only beyond the expressive power of equations, but, by Gödel's incompleteness theorem, is undefinable even in first-order logic.

Remarkably enough, up to categorical equivalence, lattice-ordered abelian groups with a strong unit can be defined by equations—the equations of MV-algebras. Since these equations are nothing more than a reformulation of tautologies in the infinite-valued calculus, and since the infinite-valued tautology problem is no more complicated than its boolean counterpart, it becomes natural to apply to other mathematical areas logic-algorithmic notions originating from the many-valued calculus. For this purpose, one may use lattice-ordered abelian groups with strong unit, via the Γ functor, as a bridge between MV-algebras and other structures. Important examples are given by approximately finite-dimensional (AF) C^*-algebras, the algebras of operators currently used for the mathematical description of infinite spin systems. As another interesting example, disjunctive normal form reductions for formulas in the infinite-valued calculus are essentially the same as desingularization algorithms for toric varieties, once the latter are described by their associated fans—a fan being a complex of rational polyhedral cones.

The relationships between MV-algebras, C^*-algebras, and toric varieties will be briefly discussed in Chapter 9; most of that chapter can be safely skipped by readers only interested in the Łukasiewicz calculus and its algebras. Similarly, we have made no attempt to introduce here "first-order" infinite-valued logic: as a matter of fact, the development of the infinite-valued counterparts of such notions as "set", "equality", "structure" would result in a much less elementary textbook of considerably larger size—let alone the problem of choosing the right definitions. In a final section, we give appropriate bibliographical references to the interested reader.

While it would be beyond the scope of the book to cover all the fascinating and rapidly developing fields of research connected with infinite-valued reasoning, for most of these topics the book is intended to provide sufficient background material. Thus, Chapter 8 is devoted to the classification of equational classes of MV-algebras; these include all classes associated to the finite-valued calculi of Łukasiewicz. The fruitful interplay between MV-algebras and lattices is discussed at length in Chapter 6. Extensive bibliographical references are given in the final chapter on Further Readings, concerning such basic issues as the "multiplication connective", as well as probability and proof theory in the infinite-valued propositional calculus of Łukasiewicz.

* * *

While the first germ of this book is our monograph [58], the proofs of many fundamental theorems are given here in a more general and self-contained form, using results that have appeared in the literature after the publication of [58]. Several sections on advanced topics have been added, and the bibliography has been considerably expanded. The book also contains a wealth of previously unpublished material.

This book is didactic in its spirit: preliminary versions have been tested in several graduate courses in Bahía Blanca, Barcelona, Buenos Aires, Campinas, Mérida, Milan, Patras, and in the general context of the European Project known as Action COST number 15 on "Many-valued Logic for Computer Science Applications". We are grateful to Stefano Aguzzoli, Agata Ciabattoni, Vincenzo Marra and Claudia Picardi for their valuable comments on earlier drafts of this book. For valuable hints and discussions we are indebted to many more students and colleagues: we ask them to forgive us for not listing them here. We also thank the anonymous referee for his careful and competent reading and for suggesting several improvements.

We gratefully acknowledge partial support from the National Research Councils of Argentina (CONICET), Brazil (CNPq) and Italy

(CNR), as well as from "Fundación Antorchas" (Buenos Aires), "Centre de Recerca Matemática, Institut d'Estudis Catalans" (Barcelona), Fundação de Amparo á Pesquisa do Estado de São Paulo (FAPESP, São Paulo), Istituto Italiano per gli Studi Filosofici (Naples), besides our own universities, and COST Action 15. Their support allowed us to meet several times in Argentina, Brazil and Italy.

R.C., I.M.L.D.O, D.M.

Chapter 1

Basic notions

We introduce MV-algebras by means of a small number of simple equations, in an attempt to capture certain properties of the unit real interval $[0, 1]$ equipped with truncated addition $x \oplus y = \min(1, x + y)$ and negation $1 - x$. We show that every MV-algebra contains a natural lattice-order. The chapter culminates with Chang's Subdirect Representation Theorem, stating that if an equation holds in all totally ordered MV-algebras, then the equation holds in all MV-algebras.

1.1 MV-algebras

Definition 1.1.1 An *MV-algebra* is an algebra $\langle A, \oplus, \neg, 0\rangle$ with a binary operation $\oplus$, a unary operation $\neg$ and a constant 0 satisfying the following equations:

MV1) $x \oplus (y \oplus z) = (x \oplus y) \oplus z$

MV2) $x \oplus y = y \oplus x$

MV3) $x \oplus 0 = x$

MV4) $\neg\neg x = x$

MV5) $x \oplus \neg 0 = \neg 0$

MV6) $\neg(\neg x \oplus y) \oplus y = \neg(\neg y \oplus x) \oplus x.$

In particular, axioms MV1)–MV3) state that $\langle A, \oplus, 0\rangle$ is an *abelian monoid*. Following tradition, we denote an MV-algebra $\langle A, \oplus, \neg, 0\rangle$ by its universe A.

A singleton $\{0\}$ is a trivial example of an MV-algebra. An MV-algebra is said *nontrivial* iff its universe has more than one element. As a first example of a nontrivial MV-algebra, consider the real unit interval $[0,1] = \{x \in \mathbf{R} \mid 0 \leq x \leq 1\}$, and for all $x, y \in [0,1]$, let $x \oplus y =_{def} \min(1, x+y)$ and $\neg x =_{def} 1 - x$. It is easy to see that $[0,1] =_{def} \langle [0,1], \oplus, \neg, 0\rangle$ is an MV-algebra. As a second example, if $\langle A, \vee, \wedge, -, 0, 1\rangle$ is a boolean algebra, then $\langle A, \vee, -, 0\rangle$ is an MV-algebra, where $\vee$, $-$ and 0 denote, respectively, the join, the complement and the smallest element in A.

A *subalgebra* of an MV-algebra A is a subset S of A containing the zero element of A, closed under the operations of A, and equipped with the restriction to S of these operations.

The intersection of any nonempty family of subalgebras of A is a subalgebra of A. Given a subset X of A the intersection of all subalgebras of A containing X is a subalgebra of A, and is called the *subalgebra of A generated by X*.

The rational numbers in $[0,1]$, and, for each integer $n \geq 2$, the n-element set

$$\mathbf{L}_n =_{def} \{0, 1/(n-1), \ldots, (n-2)/(n-1), 1\},$$

yield examples of subalgebras of $[0,1]$.

Given an MV-algebra A and a set X, the set A^X of all functions $f\colon X \to A$ becomes an MV-algebra if the operations $\oplus$ and $\neg$ and the element 0 are defined pointwise. The continuous functions from $[0,1]$ into $[0,1]$ form a subalgebra of the MV-algebra $[0,1]^{[0,1]}$.

On each MV-algebra A we define the constant 1 and the operations $\odot$ and $\ominus$ as follows:

(1.1) $1 =_{def} \neg 0,$

(1.2) $x \odot y =_{def} \neg(\neg x \oplus \neg y),$

(1.3) $x \ominus y =_{def} x \odot \neg y.$

An MV-algebra is nontrivial if and only if $0 \neq 1$. The following identities are immediate consequences of MV4):

MV7) $\neg 1 = 0$, and

MV8) $x \oplus y = \neg(\neg x \odot \neg y)$.

Axioms MV5) and MV6) can now be written as:

MV5′) $x \oplus 1 = 1$, and

MV6′) $(x \ominus y) \oplus y = (y \ominus x) \oplus x$.

Setting $y = \neg 0$ in MV6) we obtain:

MV9) $x \oplus \neg x = 1$.

Note that in the MV-algebra $[0, 1]$ we have $x \odot y = \max(0, x + y - 1)$ and $x \ominus y = \max(0, x - y)$.

Notation: Following common usage, we consider the $\neg$ operation more binding than any other operation, and the $\odot$ operation more binding than $\oplus$ and $\ominus$.

Lemma 1.1.2 *Let A be an MV-algebra and x, $y \in A$. Then the following conditions are equivalent:*

(i) $\neg x \oplus y = 1$;

(ii) $x \odot \neg y = 0$;

(iii) $y = x \oplus (y \ominus x)$;

(iv) there is an element $z \in A$ such that $x \oplus z = y$.

Proof: (i) $\Rightarrow$ (ii) By MV4) and MV7). (ii) $\Rightarrow$ (iii) Immediate from MV3) and MV6′). (iii) $\Rightarrow$ (iv) Take $z = y \ominus x$. (iv) $\Rightarrow$ (i) By MV9), $\neg x \oplus x \oplus z = 1$. □

Let A be an MV-algebra. For any two elements x and y of A let us agree to write

$$x \leq y$$

iff x and y satisfy the above equivalent conditions (i)-(iv). It follows that $\leq$ is a partial order, called the *natural order* of A. Indeed, reflexivity is equivalent to MV9), antisymmetry follows from conditions (ii) and (iii), and transitivity follows from condition (iv).

An MV-algebra whose natural order is total is called an *MV-chain.* Note that, by (iv), the natural order of the MV-chain $[0,1]$ coincides with the natural order of the real numbers.

Lemma 1.1.3 *Let A be an MV-algebra. For each $a \in A$, $\neg a$ is the unique solution x of the simultaneous equations:*

$$(1.4) \quad \begin{cases} a \oplus x &= 1 \\ a \odot x &= 0 \end{cases}$$

Proof: By Lemma 1.1.2, these two equations amount to writing $\neg a \leq x \leq \neg a$. □

Lemma 1.1.4 *In every MV-algebra A the natural order $\leq$ has the following properties:*

(i) $x \leq y$ *if and only if* $\neg y \leq \neg x$;

(ii) If $x \leq y$ *then for each* $z \in A$, $x \oplus z \leq y \oplus z$ *and* $x \odot z \leq y \odot z$;

(iii) $x \odot y \leq z$ *iff* $x \leq \neg y \oplus z$.

Proof: (i) This follows from Lemma 1.1.2(i), since $\neg x \oplus y = \neg\neg y \oplus \neg x$. (ii) The monotonicity of $\oplus$ is an easy consequence of Lemma 1.1.2(iv); using (i), one immediately proves the monotonicity of $\odot$. (iii) It is sufficient to note that $x \odot y \leq z$ is equivalent to $1 = \neg(x \odot y) \oplus z = \neg x \oplus \neg y \oplus z$. □

Proposition 1.1.5 *On each MV-algebra A the natural order determines a* lattice *structure. Specifically, the join $x \vee y$ and the meet $x \wedge y$ of the elements x and y are given by*

(1.5) $x \vee y = (x \odot \neg y) \oplus y = (x \ominus y) \oplus y$,

(1.6) $x \wedge y = \neg(\neg x \vee \neg y) = x \odot (\neg x \oplus y)$.

Proof: To prove (1.5), by MV6′), MV9) and Lemma 1.1.4(ii), $x \leq (x \ominus y) \oplus y$ and $y \leq (x \ominus y) \oplus y$. Suppose $x \leq z$ and $y \leq z$. By (i) and (iii) in Lemma 1.1.2, $\neg x \oplus z = 1$ and $z = (z \ominus y) \oplus y$. Then by MV6′) we can write

$$\neg((x \ominus y) \oplus y) \oplus z = (\neg(x \ominus y) \ominus y) \oplus y \oplus (z \ominus y)$$

$$= (y \ominus \neg(x \ominus y)) \oplus \neg(x \ominus y) \oplus (z \ominus y)$$

$$= (y \ominus \neg(x \ominus y)) \oplus \neg x \oplus y \oplus (z \ominus y)$$

$$= (y \ominus \neg(x \ominus y)) \oplus \neg x \oplus z = 1.$$

It follows that $(x \ominus y) \oplus y \leq z$, which completes the proof of (1.5). We now immediately obtain (1.6) as a consequence of (1.5) together with Lemma 1.1.4(i). □

Proposition 1.1.6 *The following equations hold in every MV-algebra:*

(i) $x \odot (y \vee z) = (x \odot y) \vee (x \odot z)$,

(ii) $x \oplus (y \wedge z) = (x \oplus y) \wedge (x \oplus z)$.

Proof: By MV6′) and Lemma 1.1.4(ii), $x \odot y \leq x \odot (y \vee z)$ and $x \odot z \leq x \odot (y \vee z)$. Suppose $x \odot y \leq t$ and $x \odot z \leq t$. Then by Lemma 1.1.4(iii), $y \leq \neg x \oplus t$ and $z \leq \neg x \oplus t$, whence $y \vee z \leq \neg x \oplus t$. One more application of Lemma 1.1.4 (iii) yields $(y \vee z) \odot x \leq t$, which completes the proof of (i). It is now easy to see that (ii) is a consequence of (i), using Lemma 1.1.4(i), together with MV4) and MV8). □

Proposition 1.1.7 *Every MV-algebra satisfies the equation*

(1.7) $(x \ominus y) \wedge (y \ominus x) = 0$.

Proof: By making repeated use of MV6) and its variants, together with the basic properties of the operations $\oplus$ and $\odot$ we obtain:

$$(x \ominus y) \wedge (y \ominus x)$$

$$= (x \ominus y) \odot (\neg(x \ominus y) \oplus (y \ominus x))$$

$$= x \odot \neg y \odot (y \oplus \neg x \oplus (y \ominus x))$$

$$= x \odot (\neg x \oplus (y \ominus x)) \odot (\neg(\neg x \oplus (y \ominus x)) \oplus \neg y)$$

$$= (y \ominus x) \odot (\neg(y \ominus x) \oplus x) \odot (\neg(\neg x \oplus (y \ominus x)) \oplus \neg y)$$

$$= y \odot \neg x \odot (\neg(y \ominus x) \oplus x) \odot ((x \odot \neg(y \ominus x)) \oplus \neg y)$$

$$= \neg x \odot (x \oplus \neg(y \ominus x)) \odot y \odot (\neg y \oplus (x \odot (\neg y \oplus x))) =$$

$$\neg x \odot (x \oplus \neg(y \ominus x)) \odot (x \odot (\neg y \oplus x)) \odot (\neg(x \odot (\neg y \oplus x)) \oplus y)$$

$= 0$, since by MV8) and MV9), $\neg x \odot x = 0$. □

Let A be an MV-algebra. For each $x \in A$, we let $0x = 0$, and for each integer $n \geq 0$, $(n+1)x = nx \oplus x$.

Lemma 1.1.8 *Let x and y be elements of an MV-algebra A. If $x \wedge y = 0$ then for each integer $n \geq 0$, $nx \wedge ny = 0$.*

Proof: If $x \wedge y = 0$ then by monotonicity (Lemma 1.1.4) and distributivity (Proposition 1.1.6), $x = x \oplus (x \wedge y) = (x \oplus x) \wedge (x \oplus y) \geq 2x \wedge y$, whence $0 = x \wedge y \geq 2x \wedge y$. It follows that $0 = 2x \wedge 2y = 4x \wedge 4y = 8x \wedge 8y = \ldots$. The desired conclusion now follows from $nx \wedge ny \leq 2^n x \wedge 2^n y = 0$. □

1.2 Homomorphisms and ideals

Let A and B be MV-algebras. A function $h: A \to B$ is a *homomorphism* iff it satisfies the following conditions, for each $x, y \in A$:

H1) $h(0) = 0$,

H2) $h(x \oplus y) = h(x) \oplus h(y)$,

H3) $h(\neg x) = \neg h(x)$.

Following current usage, if h is one-one we shall equivalently say that h is an *injective* homomorphism, or an *embedding*. If the homomorphism $h: A \to B$ is onto B we say that h is *surjective*. By an *isomorphism* we shall mean a surjective one-one homomorphism. We write $A \cong B$ iff there is an isomorphism from A onto B.

The *kernel* of a homomorphism $h: A \to B$ is the set

$$Ker(h) =_{def} h^{-1}(0) = \{x \in A \mid h(x) = 0\}.$$

Our next aim is to characterize kernels of homomorphisms.

An *ideal* of an MV-algebra A is a subset I of A satisfying the following conditions:

I1) $0 \in I$,

I2) If $x \in I$, $y \in A$ and $y \leq x$ then $y \in I$,

I3) If $x \in I$ and $y \in I$ then $x \oplus y \in I$.

The intersection of any family of ideals of A is still an ideal of A. For every subset $W \subseteq A$, the intersection of all ideals $I \supseteq W$ is said to be the ideal *generated* by W, and will be denoted $\langle W \rangle$.

The proof of the next lemma is immediate, and will be omitted.

Lemma 1.2.1 *Let W be a subset of an MV-algebra A. If $W = \emptyset$, then $\langle W \rangle = \{0\}$. If $W \neq \emptyset$, then*

$$\langle W \rangle = \{x \in A \mid x \leq w_1 \oplus \cdots \oplus w_k, \text{ for some } w_1, \ldots, w_k \in W\}.$$

□

In particular, for each element z of an MV-algebra A, the ideal $\langle z \rangle = \langle \{z\} \rangle$ is called the *principal ideal generated by* z, and we have

(1.8) $\langle z \rangle = \{x \in A \mid nz \geq x$ for some integer $n \geq 0\}$.

Note that $\langle 0 \rangle = \{0\}$ and $\langle 1 \rangle = A$. Further, for every ideal J of an MV-algebra A and each $z \in A$ we have

(1.9) $\langle J \cup \{z\} \rangle = \{x \in A \mid x \leq nz \oplus a$, for some $n \in \mathbf{N}$ and $a \in J\}$.

An ideal I of an MV-algebra A is *proper* iff $I \neq A$. We say that I is *prime* iff it is proper and satisfies the following condition:

I4) For each x and y in A, either $(x \ominus y) \in I$ or $(y \ominus x) \in I$.

An ideal I of an MV-algebra A is called *maximal* iff it is proper and no proper ideal of A strictly contains I, i.e., for each ideal $J \neq I$, if $I \subseteq J$ then $J = A$.

The next proposition generalizes a well known property of maximal ideals in boolean algebras.

Proposition 1.2.2 *For any proper ideal J of an MV-algebra A the following conditions are equivalent:*

(i) J is a maximal ideal of A;

(ii) for each $x \in A$, $x \notin J$ iff $\neg nx \in J$ for some integer $n \geq 1$.

Proof: (i) $\rightarrow$ (ii): Suppose that J is a maximal ideal of A. If $x \notin J$, then $\langle\{x\} \cup J\rangle = A$, and by (1.9), $1 = nx \oplus a$ for some integer $n \geq 1$ and $a \in J$. Stated otherwise, $\neg nx \leq a \in J$, whence by **I2)**, $\neg nx \in J$. Conversely, if $x \in J$, then $nx \in J$ for each integer $n \geq 1$; since J is proper, $\neg nx \notin J$.

(ii) $\rightarrow$ (i): Let $K \neq J$ be an ideal of A such that $J \subseteq K$. For every $x \in K \setminus J$ we must have $\neg nx \in J$ for some integer $n \geq 1$. Hence $1 = nx \oplus \neg nx \in K$, and $K = A$. □

We denote by $\mathcal{I}(A)$, $\mathcal{P}(A)$ and $\mathcal{M}(A)$ the sets of ideals, prime ideals and maximal ideals of A, respectively.

In the next lemma we summarize, for further reference, some easy relations between ideals and kernels of homomorphisms.

Lemma 1.2.3 *Let A, B be MV-algebras, and $h: A \rightarrow B$ a homomorphism. Then the following properties hold:*

(i) For each ideal J of B, the set $h^{-1}(J) =_{def} \{x \in A \mid h(x) \in J\}$ is an ideal of A. Thus in particular, $Ker(h) \in \mathcal{I}(A)$;

(ii) $h(x) \leq h(y)$ iff $x \ominus y \in Ker(h)$;

(iii) h is injective iff $Ker(h) = \{0\}$;

(iv) $Ker(h) \neq A$ iff B is nontrivial;

(*v*) *$Ker(h) \in \mathcal{P}(A)$ iff B is nontrivial and the image $h(A)$, as a subalgebra of B, is an MV-chain.* □

Definition 1.2.4 The *distance function* $d\colon A \times A \longrightarrow A$ is defined by

(1.10) $d(x,y) =_{def} (x \ominus y) \oplus (y \ominus x)$.

In the MV-algebra $[0,1]$, $d(x,y) = |x-y|$. In every boolean algebra the distance function coincides with the symmetric difference operation.

Proposition 1.2.5 *In every MV-algebra A we have:*

(*i*) $d(x,y) = 0$ *iff* $x = y$,

(*ii*) $d(x,y) = d(y,x)$,

(*iii*) $d(x,z) \leq d(x,y) \oplus d(y,z)$,

(*iv*) $d(x,y) = d(\neg x, \neg y)$,

(*v*) $d(x \oplus s, y \oplus t) \leq d(x,y) \oplus d(s,t)$.

Proof: Properties (i), (ii) and (iv) immediately follow by definition, recalling the basic properties of the natural order on A (Lemmas 1.1.2 and 1.1.4). To prove (iii), note first that $\neg(x \ominus z) \oplus (x \ominus y) \oplus (y \ominus z) = (\neg x \vee \neg y) \oplus (z \vee y) \geq \neg y \oplus y = 1$. Hence, $(x \ominus z) \leq (x \ominus y) \oplus (y \ominus z)$. In a similar way we obtain $(z \ominus x) \leq (y \ominus x) \oplus (z \ominus y)$, whence (iii) follows from the monotonicity of $\oplus$ (Lemma 1.1.4(ii)). One similarly proves (v) by observing that $\neg((x \oplus s) \ominus (y \oplus t)) \oplus (x \ominus y) \oplus (s \ominus t) = \neg(x \oplus s) \oplus (x \vee y) \oplus (t \vee s) \geq \neg(x \oplus s) \oplus x \oplus s = 1$. □

As an immediate consequence we have

Proposition 1.2.6 *Let I be an ideal of an MV-algebra A. Then the binary relation $\equiv_I$ on A defined by $x \equiv_I y$ iff $d(x,y) \in I$ is a* congruence *relation. (Stated otherwise, $\equiv_I$ is an equivalence relation such that $x \equiv_I s$ and $y \equiv_I t$ imply $\neg x \equiv_I \neg s$ and $x \oplus y \equiv_I s \oplus t$.) Moreover, $I = \{x \in A \mid x \equiv_I 0\}$.*

Conversely, if $\equiv$ is a congruence on A, then $\{x \in A \mid x \equiv 0\}$ is an ideal, and $x \equiv y$ iff $d(x,y) \equiv 0$. Therefore, the correspondence $I \mapsto \equiv_I$ is a bijection from the set of ideals of A onto the set of congruences on A. □

Given $x \in A$, the equivalence class of x with respect to $\equiv_I$ will be denoted by x/I and the quotient set $A/\equiv_I$ by A/I. Since $\equiv_I$ is a congruence, defining on the set A/I the operations

(1.11) $\neg(x/I) =_{def} \neg x/I$

and

(1.12) $x/I \oplus y/I =_{def} (x \oplus y)/I,$

the system $\langle A/I, \oplus, \neg, 0/I \rangle$ becomes an MV-algebra, called *the quotient algebra of A by the ideal I.* Moreover, the correspondence $x \mapsto x/I$ defines a homomorphism h_I from A onto the quotient algebra A/I, which is called *the natural homomorphism from A onto A/I*. Note that $Ker(h_I) = I$.

The next lemma is an easy consequence of Lemma 1.2.3(ii).

Lemma 1.2.7 *If A, B and C are MV-algebras, and $f: A \rightarrow B$ and $g: A \rightarrow C$ are surjective homomorphisms, then $Ker(f) \subseteq Ker(g)$ if and only if there is a surjective homomorphism $h: B \rightarrow C$ such that $h \circ f = g$, i.e., $h(f(x)) = g(x)$ for all $x \in A$. This homomorphism h is an isomorphism if and only if $Ker(f) = Ker(g)$.* □

Upon noting that $Ker(h) = Ker(h_{Ker(h)})$ we immediately get

Theorem 1.2.8 *Let A and B be MV-algebras. If $h: A \rightarrow B$ is a surjective homomorphism, then there is an isomorphism $f: A/Ker(h) \rightarrow B$ such that $f(x/Ker(h)) = h(x)$ for all $x \in A$.* □

Proposition 1.2.9 *If A is an MV-chain, then all proper ideals of A are prime.*

Proof: Let I be a proper ideal of A. Since $h_I: A \rightarrow A/I$ is a surjective homomorphism, then A/I is also an MV-chain, and hence, by Lemma 1.2.3(v), I must be a prime ideal. □

Proposition 1.2.10 *Let J be an ideal of an MV-algebra A. Then the map $I \mapsto h_J(I)$ determines an inclusion preserving one-one correspondence between the ideals of A containing J and the ideals of the quotient MV-algebra A/J. The inverse map also preserves inclusions, and is obtained by taking the inverse image $h_J^{-1}(K)$ of each ideal K of A/J.*

Proof: Let I be an ideal of A such that $J \subseteq I$. Since h_J maps A onto A/J and $Ker(h_J) = J \subseteq I$, by Lemma 1.2.3 (ii) and (MV6′), we have $h_J(I) \in \mathcal{I}(A/J)$ and, moreover, $h_J^{-1}(h_J(I)) \subseteq I$. Since the converse inclusion holds for all surjective mappings, one has that $I = h_J^{-1}(h_J(I))$. On the other hand, by Lemma 1.2.3 (i), $h_J^{-1}(K) \in \mathcal{I}(A)$ for each $K \in \mathcal{I}(A/J)$. To complete the proof it is sufficient to note that $J = h_J^{-1}(\{0\}) \subseteq h_J^{-1}(K)$ and $h_J(h_J^{-1}(K)) = K$. □

Remark: If A is an MV-chain, then the set $\mathcal{I}(A)$ is totally ordered by inclusion. Indeed, if I, J were ideals of A such that $I \not\subseteq J$ and $J \not\subseteq I$, then there would be elements a, b in A such that $a \in I \setminus J$ and $b \in J \setminus I$, whence $a \not\leq b$ and $b \not\leq a$, which is impossible.

Theorem 1.2.11 *The following properties hold in any MV-algebra A:*

(*i*) *Every proper ideal of A that contains a prime ideal is prime;*

(*ii*) *For each prime ideal J of A, the set $\{I \in \mathcal{I}(A) \mid J \subseteq I\}$ is totally ordered by inclusion.*

Proof: Let J be a prime ideal of A. By Lemma 1.2.3(v), A/J is an MV-chain, and by Proposition 1.2.9 and the above remark, all proper ideals of A/J are prime and are totally ordered by inclusion. This, together with Proposition 1.2.10, implies (ii). To prove (i), note that if I is a proper ideal of A such that $J \subseteq I$, then again by Proposition 1.2.10, $I = h_J^{-1}(h_J(I))$, whence I is the inverse image of a prime ideal of A/J. □

Corollary 1.2.12 *Every prime ideal J of an MV-algebra A is contained in a unique maximal ideal of A.*

Proof: The set $\mathcal{H} =_{def} \{I \in \mathcal{I}(A) \mid I \neq A \text{ and } J \subseteq I\}$ is totally ordered by inclusion. Therefore, $M =_{def} \bigcup_{I \in \mathcal{H}} I$ is an ideal of A. Further, M is a proper ideal, because $1 \notin M$; we conclude that M is the only maximal ideal containing J. □

The next proposition will play an important role in the proof of Chang's Subdirect Representation Theorem.

Proposition 1.2.13 *Let A be an MV-algebra, J an ideal of A, and $a \in A \setminus J$. Then there is a prime ideal P of A such that $J \subseteq P$ and $a \notin P$.*

Proof: A routine application of Zorn's Lemma shows that there is an ideal I of A which is maximal with respect to the property that $J \subseteq I$ and $a \notin I$. We shall show that I is a prime ideal. Let x and y be elements of A, and suppose that both $x \ominus y \notin I$ and $y \ominus x \notin I$ (absurdum hypothesis). Then the ideal generated by I and $x \ominus y$ must contain the element a. By (1.9), $a \leq s \oplus p(x \ominus y)$ for some $s \in I$ and some integer $p \geq 1$. Similarly, there is an element $t \in I$ and an integer $q \geq 1$ such that $a \leq t \oplus q(y \ominus x)$. Let $u = s \oplus t$ and $n = \max(p, q)$. Then $u \in I$, $a \leq u \oplus n(x \ominus y)$ and $a \leq u \oplus n(y \ominus x)$. Hence by (1.6) and (1.7), together with Proposition 1.1.6(ii) and Lemma 1.1.8, we have $a \leq (u \oplus n(x \ominus y)) \wedge (u \oplus n(y \ominus x)) = u \oplus (n(x \ominus y) \wedge n(y \ominus x))$ $= u$, whence $a \in I$, a contradiction. □

Corollary 1.2.14 *Every proper ideal of an MV-algebra is an intersection of prime ideals.* □

From the above proposition and Corollary 1.2.12 we immediately obtain:

Corollary 1.2.15 *Every nontrivial MV-algebra has a maximal ideal.* □

In the next proposition we generalize some properties of maximal ideals of boolean algebras. Remarkably enough, (the analogues of) these properties do not hold for other extensions of boolean algebras, such as bounded distributive lattices or commutative rings with unit.

Proposition 1.2.16 *Let A and B be MV-algebras, and M be a maximal ideal of B. Then we have*

(i) For any homomorphism $h: A \to B$, the inverse image $h^{-1}(M)$ is a maximal ideal of A;

(ii) For any subalgebra S of B, $S \cap M$ is a maximal ideal of S.

Proof: (i) By Lemma 1.2.3, $h^{-1}(M)$ is an ideal of A; since $h(1) = 1 \notin M$, then $h^{-1}(M)$ must be a proper ideal. Suppose $z \notin h^{-1}(M)$. Since $h(z) \notin M$, then by Proposition 1.2.2 there is an integer $n \geq 1$ such that $\neg nh(z) \in M$. It follows that $\neg nz \in h^{-1}(M)$, whence, again by Proposition 1.2.2, $h^{-1}(M)$ is a maximal ideal of A.

The proof of (ii) easily follows from (i), upon letting $\iota: S \to B$ be the natural embedding, given by $\iota(x) = x$ for all $x \in S$, and noting that $M \cap B = \iota^{-1}(M)$. □

1.3 Subdirect representation theorem

Throughout this section I shall denote a nonempty set. The *direct product* of a family $\{A_i\}_{i \in I}$ of MV-algebras, denoted by $\prod_{i \in I} A_i$, is the MV-algebra obtained by endowing the set-theoretical cartesian product of the family with the MV-operations defined pointwise. In other words, $\prod_{i \in I} A_i$ is the set of all functions $f: I \to \bigcup_{i \in I} A_i$ such that $f(i) \in A_i$ for all $i \in I$, with the operations $\neg$ and $\oplus$ defined by

$$(\neg f)(i) =_{def} \neg f(i) \quad \text{and} \quad (f \oplus g)(i) =_{def} f(i) \oplus g(i).$$

The zero element of $\prod_{i \in I} A_i$ is the function $i \in I \mapsto 0_i \in A_i$. For each $j \in I$, the map $\pi_j: \prod_{i \in I} A_i \to A_j$ is defined by

$$\pi_j(f) =_{def} f(j).$$

Each π_j is a homomorphism onto A_j, called the jth *projection function.* In particular, for each MV-algebra A and nonempty set X, the MV-algebra A^X is the direct product of the family $\{A_x\}_{x \in X}$, where $A_x = A$ for all $x \in X$.

Definition 1.3.1 An MV-algebra A is a *subdirect product* of a family $\{A_i\}_{i \in I}$ of MV-algebras iff there exists a one-one homomorphism $h: A \to \prod_{i \in I} A_i$ such that for each $j \in I$, the composite map $\pi_j \circ h$ is a homomorphism *onto* A_j.

If A is a subdirect product of the family $\{A_i\}_{i \in I}$, then A is isomorphic to the subalgebra $h(A)$ of $\prod_{i \in I} A_i$; moreover, the restriction to $h(A)$ of each projection is a surjective mapping.

The following result is a particular case of a theorem of Universal Algebra, due to Birkhoff:

Theorem 1.3.2 *An MV-algebra A is a subdirect product of a family $\{A_i\}_{i\in I}$ of MV-algebras if and only if there is a family $\{J_i\}_{i\in I}$ of ideals of A such that*

(i) $A_i \cong A/J_i$ *for each* $i \in I$

and

(ii) $\bigcap_{i\in I} J_i = \{0\}$.

Proof: Supposing first that A is a subdirect product of a family $\{A_i\}_{i\in I}$ of MV-algebras, let $h: A \to \prod_{i\in I} A_i$ be a one-one homomorphism as given by Definition 1.3.1; for each $j \in I$, let $J_j = Ker(\pi_j \circ h)$. By Theorem 1.2.8, $A_j \cong A/J_j$. If $x \in \bigcap_{i\in I} J_i$, then $\pi_j(h(x)) = 0$ for all $j \in I$. This implies that $h(x) = 0$, and since h is injective, $x = 0$. Therefore $\bigcap_{i\in I} J_i = \{0\}$, and conditions (i) and (ii) hold true.

Conversely, suppose $\{J_i\}_{i\in I}$ to be a family of ideals of A satisfying conditions (i) and (ii). Let ϵ_i be an isomorphism of A/J_i onto A_i, as given by condition (i). Let the function $h: A \to \prod_{i\in I} A_i$ be defined by stipulating that, for each $x \in A$, $(h(x))(i) = \epsilon_i(x/J_i)$. It follows from (ii) that $Ker(h) = \{0\}$, whence, by Lemma 1.2.3(iii), h is injective. Since for each $i \in I$ the map $a \in A \mapsto a/J_i \in A/J_i$ is surjective, then $\pi_i \circ h$ maps A onto A_i. Thus, A is a subdirect product of the family $\{A_i\}_{i\in I}$, as required. □

The following result is fundamental:

Theorem 1.3.3 (Chang's Subdirect Representation Theorem) *Every nontrivial MV-algebra is a subdirect product of MV-chains.*

Proof: By Theorem 1.3.2 and Lemma 1.2.3(v), an MV-algebra A is a subdirect product of a family of MV-chains if and only if there is a family $\{P_i\}_{i\in I}$ of prime ideals of A such that $\bigcap_{i\in I} P_i = \{0\}$. Now apply Corollary 1.2.14 to the ideal $\{0\}$. □

1.4 MV-equations

As we shall see, an important consequence of Chang's Subdirect Representation Theorem is that in order to prove that an equation holds in all MV-algebras it is sufficient to check that the equation holds in all MV-chains. To give a precise formulation to this result we shall now develop the necessary syntactic machinery.

Definition 1.4.1 By a *string* (or, *word*) over a nonempty set S we understand a finite list of elements of S. The latter are often called the *symbols* of *alphabet* S.

For each natural number $t \geq 1$, let $S_t =_{def} \{0, \neg, \oplus, x_1, \ldots, x_t, (,)\}$. An *MV-term* in the variables $x_1, \ldots, x_t$ is a string over S_t arising from a finite number of applications of the following rules:

(T1) The elements 0 and x_i, for $i = 1, \ldots, t$, considered as one-element strings, are MV-terms.

(T2) If the string τ is an MV-term, then so is $\neg\tau$.

(T3) If the strings τ and σ are MV-terms, then so is $(\tau \oplus \sigma)$.

In other words, a string τ over S_t is an MV-term if and only if there is a *formation (or, parsing) sequence for* τ, i.e., a finite list of strings over S_t, say $\tau_1, \ldots, \tau_n$, such that $\tau_n = \tau$ and for each $i \in \{1, \ldots, n\}$, τ_i satisfies *at least one* of the following conditions:

(i) $\tau_i = 0$ or $\tau_i = x_j$, for some $1 \leq j \leq t$,

(ii) there is $j < i$ such that $\tau_i = \neg\tau_j$,

(iii) there are $j < i$ and $k < i$ such that $\tau_i = (\tau_j \oplus \tau_k)$.

Those strings τ_i that belong to every formation sequence for τ are said to be the *subterms* of τ.

The following result is known as the *unique readability* theorem; its proof is precisely the same as for the classical propositional calculus, and is left as a routine exercise.

Theorem 1.4.2 *Every term τ_i in the variables $x_1, \ldots, x_n$ satisfies* precisely one *of the above conditions (i)-(iii). Moreover, both term τ_j of case (ii) and the pair (τ_j, τ_k) of case (iii) are uniquely determined.* □

We shall henceforth write $\tau(x_1, \ldots, x_n)$ to signify that τ is an MV-term in the variables $x_1, \ldots, x_n$.

The following strings are examples of MV-terms in the variables x_1, x_2, x_3:

$$(\neg 0 \oplus x_1), ((x_2 \oplus 0) \oplus \neg(x_3 \oplus x_1)) \text{ and } (x_2 \oplus \neg x_1).$$

Notation: Following tradition, for the sake of readability, in writing terms we shall omit the outermost pair of brackets. Further, we shall freely use the symbols $\odot$, $\ominus$, $\vee$, $\wedge$ and 1 to write MV-terms in abbreviated form, in the light of (1.1)-(1.6). Thus, given MV-terms τ and σ in the variables $x_1, \ldots, x_t$, we shall freely call MV-terms (in the same variables) also such abbreviations as $\tau \odot \sigma$, $\tau \ominus \sigma$, $\tau \vee \sigma$, $\tau \wedge \sigma$, and the like.

Definition 1.4.3 Let A be an MV-algebra, τ an MV-term in the variables $x_1, \ldots, x_t$, and assume $a_1, \ldots, a_t$ are elements of A. Substituting an element $a_i \in A$ for all occurrences of the variable x_i in τ, for $i = 1, \ldots, t$, using the unique readability theorem, and interpreting the symbols 0, $\oplus$ and $\neg$ as the corresponding operations in A, we obtain an element of A, denoted $\tau^A(a_1, \ldots, a_t)$.

In more detail, proceeding by induction on the number of operation symbols occurring in τ, we define $\tau^A(a_1, \ldots, a_t)$ as follows:

(i) $x_i^A = a_i$, for each $i = 1, \ldots, t$;

(ii) $(\neg\sigma)^A = \neg(\sigma^A)$;

(iii) $(\sigma \oplus \rho)^A = (\sigma^A \oplus \rho^A)$.

By the above definition, given any MV-algebra A we can associate each MV-term τ in the variables $x_1, \ldots, x_n$ with a function $\tau^A\colon A^n \to A$. Functions arising in this way are called *term functions on A*. The dependence of τ^A on n is tacitly understood.

Definition 1.4.4 An *MV-equation* (for short, an *equation*) in the variables $x_1, \ldots, x_t$ is a pair (τ, σ) of MV-terms in the variables $x_1, \ldots, x_t$.

Following tradition, we shall write $\tau = \sigma$ instead of (τ, σ). An MV-algebra A *satisfies* the MV-equation $\tau = \sigma$, in symbols,

$$A \models \tau = \sigma,$$

iff $\tau^A(a_1, \ldots, a_t) = \sigma^A(a_1, \ldots, a_t)$ for all $a_1, \ldots, a_t \in A$.

Axioms (MV1)–(MV6) are examples of MV-equations in the variables x, y, z. By definition, these equations are satisfied by all MV-algebras. In the previous sections many other equations have been shown to hold for all MV-algebras. By contrast, the equation $x \oplus x \oplus x = x \oplus x$ is satisfied by the MV-algebra $\mathbf{L}_n$ if and only if $n = 2$ or $n = 3$.

Remark: By Proposition 1.2.5(i), an MV-equation $\tau = \sigma$ holds in an MV-algebra A if and only if the equation $(\tau \ominus \sigma) \oplus (\sigma \ominus \tau) = 0$ holds in A. Therefore we can safely assume that all MV-equations are of the form $\rho = 0$, where ρ is an MV-term.

Lemma 1.4.5 *Let A, B, A_i (for all $i \in I$) be MV-algebras:*

(i) If $A \models \tau = \sigma$ then $S \models \tau = \sigma$ for each subalgebra S of A;

(ii) If $h\colon A \to B$ is a homomorphism, then for each MV-term τ in the variables $x_1, \ldots, x_s$ and each s-tuple $(a_1, \ldots, a_s)$ of elements of A we have $\tau^B(h(a_1), \ldots, h(a_s)) = h(\tau^A(a_1, \ldots, a_s))$. In particular, when h maps A onto *B, from $A \models \tau = \sigma$ it follows that $B \models \tau = \sigma$;*

(iii) If $A_i \models \tau = \sigma$ for each $i \in I$, then $\prod_{i \in I} A_i \models \tau = \sigma$.

Proof: Conditions (i) and (ii) are immediate. As for condition (iii), let $f_1, \ldots, f_s \in A = \prod_{i \in I} A_i$. By hypothesis, for each $i \in I$ we can write

$$\begin{aligned} \tau^A(f_1, \ldots, f_s)(i) &= \tau^{A_i}(f_1(i), \ldots, f_s(i)) = \\ \sigma^{A_i}(f_1(i), \ldots, f_s(i)) &= \sigma^A(f_1, \ldots, f_s)(i), \end{aligned}$$

whence $\tau^A(f_1, \ldots, f_s) = \sigma^A(f_1, \ldots, f_s)$. □

Theorem 1.4.6 *Let A be the subdirect product of a family $\{A_i\}_{i \in I}$ of MV-algebras; let $\tau = \sigma$ be an MV-equation. Then $A \models \tau = \sigma$ if and only if $A_i \models \tau = \sigma$ for each $i \in I$.*

Proof: Let $h: A \to \prod_{i\in I} A_i$ be a one-one homomorphism as given by Definition 1.3.1. Suppose that $A \models \tau = \sigma$. Since for each $i \in I$, $\pi_i \circ h$ maps A onto A_i, it follows that $A_i \models \tau = \sigma$.

Conversely, suppose that $A_i \models \tau = \sigma$ for all $i \in I$. By the above lemma, $\prod_{i\in I} A_i \models \tau = \sigma$, and since $h(A)$ is a subalgebra of $\prod_{i\in I} A_i$, $h(A) \models \tau = \sigma$. Since h^{-1} maps $h(A)$ onto A, we conclude that $A \models \tau = \sigma$. □

Corollary 1.4.7 *An MV-equation is satisfied by all MV-algebras if and only if it is satisfied by all MV-chains.*

Proof: Suppose that $\tau = \sigma$ is satisfied by all MV-chains, and let A be an MV-algebra. If $A = \{0\}$ then, trivially, $\tau^A(0,\ldots,0) = 0 = \sigma^A(0,\ldots,0)$, whence $A \models \tau = \sigma$. If A is nontrivial the desired conclusion follows from Theorems 1.3.3 and 1.4.6. □

Corollary 1.4.7 will be considerably strengthened in the next chapter, by proving that an equation holds in all MV-algebras if and only if it holds in the algebra $[0, 1]$.

1.5 Boolean algebras

We have already noted that boolean algebras are particular cases of MV-algebras. In this section we shall characterize boolean algebras among MV-algebras.

As shown in Section 1, the natural order makes every MV-algebra A into a lattice with minimum element 0 and maximum 1. We shall denote this lattice by

$$\mathbf{L}(A).$$

Recall that the lattice operations of join and meet are definable via the MV-operations by formulas (1.5) and (1.6).

A lattice is called *distributive* iff the following *distributive laws* hold:

$$x \wedge (y \vee z) = (x \wedge y) \vee (x \wedge z)$$

and

$$x \vee (y \wedge z) = (x \vee y) \wedge (x \vee z).$$

When A is an MV-chain and $a, b \in A$, then $a \vee b = \max(a, b)$ and $a \wedge b = \min(a, b)$, whence clearly the distributive laws hold in A.

Using (1.5) and (1.6), the above two distributive laws can be equivalently reformulated as MV-equations; since these equations are satisfied by all MV-chains, by Corollary 1.4.7 we obtain

Proposition 1.5.1 *For any MV-algebra A, $\mathbf{L}(A)$ is a distributive lattice with smallest element 0 and greatest element 1.* □

Definition 1.5.2 An element x of a lattice L with 0 and 1 is said to be *complemented* iff there is an element $y \in L$ (the *complement* of x) such that $x \vee y = 1$ and $x \wedge y = 0$. When L is distributive each $z \in L$ has at most one complement, denoted $-z$. We further let

$$\mathbf{B}(L)$$

be the set of all complemented elements of the distributive lattice L. Note that 0 and 1 are elements of $\mathbf{B}(L)$, because $-0 = 1$ and $-1 = 0$. As a matter of fact, $\mathbf{B}(L)$ is a sublattice of L which is also a boolean algebra. For any MV algebra A we shall write $\mathbf{B}(A)$ as an abbreviation of $\mathbf{B}(\mathbf{L}(A))$. Elements of $\mathbf{B}(A)$ are called the *boolean* elements of A.

Theorem 1.5.3 *For every element x in an MV algebra A the following conditions are equivalent:*

(i) $x \in \mathbf{B}(A)$;

(ii) $x \vee \neg x = 1$;

(iii) $x \wedge \neg x = 0$;

(iv) $x \oplus x = x$;

(v) $x \odot x = x$;

(vi) $x \oplus y = x \vee y$, *for all* $y \in A$;

(vii) $x \odot y = x \wedge y$, *for all* $y \in A$.

Proof: The following equivalences are trivial: (ii)$\Leftrightarrow$(iii), (iv)$\Leftrightarrow$(v), (vi)$\Leftrightarrow$(vii). It is also trivial that (vi) $\Rightarrow$ (iv). Further, the equivalent conditions (ii) and (iii) state that $\neg x$ is the complement of x. Thus, in particular (iii)$\Rightarrow$ (i).

(i)$\Rightarrow$ (ii): By elementary manipulations, using Lemma 1.1.2 and Proposition 1.1.6 we have $\neg x = \neg x \oplus 0 = \neg x \oplus (x \wedge -x) = (\neg x \oplus x) \wedge (\neg x \oplus -x) = \neg x \oplus -x$. Thus, $-x \leq \neg x$ and $1 = x \vee -x \leq x \vee \neg x \leq 1$, and we are done.

(iii)$\Rightarrow$(vi): Using Proposition 1.2.5, together with the Subdirect Representation Theorem 1.3.3 and the inequality $x \vee y \leq x \oplus y$, (which also is an immediate consequence of Theorem 1.3.3) we have $d(x \oplus y, x \vee y) = (x \oplus y) \odot \neg(x \vee y) = (x \oplus y) \odot (\neg x \wedge \neg y) \leq ((x \oplus y) \odot \neg x) \wedge((x \oplus y) \odot \neg y) = \neg x \wedge y \wedge \neg y \wedge x$. Therefore, $x \wedge \neg x = 0$ implies $d(x \oplus y, x \vee y) = 0$, whence $x \oplus y = x \vee y$.

(iv)$\Rightarrow$(ii): By hypothesis, $1 = \neg x \oplus x = \neg(x \oplus x) \oplus x = \neg x \vee x$. □

Corollary 1.5.4 $\mathbf{B}(A)$ *is a subalgebra of the MV algebra A. A subalgebra B of A is a boolean algebra iff* $B \subseteq \mathbf{B}(A)$. □

Corollary 1.5.5 *An MV-algebra A is a boolean algebra if and only if the operation* $\oplus$ *is* idempotent, *i.e., the equation* $x \oplus x = x$ *is satisfied by A.* □

As the reader will recall, for every element z in an MV algebra A, we denote by

(1.13) $\langle z \rangle =_{def} \{x \in A \mid x \leq z \oplus \cdots \oplus z,\ (n \text{ times}) \text{ for some } n > 0\}$

the ideal generated by z.

Corollary 1.5.6 *For any MV algebra A and* $z \in A$, *z is a boolean element of A iff the set* $\{x \in A \mid x \leq z\}$ *is an MV ideal iff* $\{x \in A \mid x \leq z\} = \langle z \rangle$.

Proof: Immediate from Theorem 1.5.3(iv). □

1.6 MV-chains

In this section we collect several results on totally ordered MV-algebras, to be used in the next chapter.

Lemma 1.6.1 *The following properties hold in every MV-chain A:*

(i) If $x \oplus y < 1$ *then* $x \odot y = 0$;

(ii) If $x \oplus y = x \oplus z$ *and* $x \odot y = x \odot z$ *then* $y = z$;

(iii) If $x \oplus y = x \oplus z < 1$ *then* $y = z$;

(iv) If $x \odot y = x \odot z > 0$ *then* $y = z$;

(v) $x \oplus y = x$ *iff* $x = 1$ *or* $y = 0$;

(vi) $x \oplus y = x$ *iff* $\neg x \oplus \neg y = \neg y$;

(vii) If $x \oplus y = 1$ *and* $x \oplus z < 1$ *then* $(x \odot y) \oplus z = (x \oplus z) \odot y$.

Proof: (i) By hypothesis, $\neg x \not\leq y$, whence $y < \neg x$. (ii) By hypothesis, $\max(\neg x, y) = \neg x \oplus (y \odot x) = \neg x \oplus (z \odot x) = \max(\neg x, z)$. Similarly, $\min(\neg x, y) = \min(\neg x, z)$, whence $y = z$. Condition (iii) is an immediate consequence of (i) and (ii). Condition (iv) follows from (iii) by Lemma 1.1.4(i). Condition (v) follows from (iii). From (v) one immediately obtains (vi). Finally, to prove (vii), since by assumption $\neg y \leq x$, we get $\neg y \oplus (x \odot y) \oplus z = (\neg y \vee x) \oplus z = x \oplus z < 1$ and $\neg y \oplus (y \odot (x \oplus z)) = \neg y \vee (x \oplus z) = x \oplus z$, whence (vii) follows from (iii). □

Remark: From Theorem 1.3.3 it follows that every MV-algebra satisfies conditions (ii) and (vi).

Proposition 1.6.2 *The following equations hold in every MV-algebra A:*

(1.14) $x \oplus y \oplus (x \odot y) = x \oplus y$

(1.15) $(x \ominus y) \oplus ((x \oplus \neg y) \odot y) = x$

(1.16) $(x \odot y) \oplus ((x \oplus y) \odot z) = (x \odot z) \oplus ((x \oplus z) \odot y)$.

Proof: By Theorem 1.3.3 we can safely assume that A is a chain. If $x \oplus y = 1$, then (1.14) follows by MV5′). If $x \oplus y < 1$, then (1.14) follows from Lemma 1.6.1(i). To prove (1.15), note that if $x \leq y$ then $x \ominus y = 0$ and $x = x \wedge y = (x \oplus \neg y) \odot y$; if, on the other hand, $y < x$ then $(x \ominus y) \oplus (x \wedge y) = (x \ominus y) \oplus y = x \vee y = x$.

As a prerequisite for the proof of (1.16) we shall prove the following equation:

$$(1.17)\quad (x \odot y) \oplus ((x \oplus y) \odot z) = (x \oplus y) \odot ((x \odot y) \oplus z).$$

Indeed, if $x \oplus y = 1$ both members of (1.17) coincide with $(x \odot y) \oplus z$. If $x \oplus y < 1$ then by Lemma 1.6.1(i) both members coincide with $(x \oplus y) \odot z$. Thus (1.17) holds for all MV-algebras.

From MV4) and MV8) we now obtain:

$$(1.18)\quad \neg((x \odot y) \oplus ((x \oplus y) \odot z)) = (\neg x \odot \neg y) \oplus ((\neg x \oplus \neg y) \odot \neg z).$$

To complete the proof of (1.16) we argue by cases as follows:

Case 1: $x \oplus y \oplus z < 1$.

Then since A is a chain, by Lemma 1.6.1(i), both members of (1.16) are equal to 0.

Case 2: $\neg x \oplus \neg y \oplus \neg z < 1$.

Same as Case 1, recalling (1.18).

There remains to consider

Case 3: $x \oplus y \oplus z = 1$ and $\neg x \oplus \neg y \oplus \neg z = 1$.

Subcase 3.1: $x \oplus y = 1$ and $x \oplus z < 1$, or $x \oplus y < 1$ and $x \oplus z = 1$.

Then by symmetry, it is sufficient to consider the case $x \oplus y = 1$ and $x \oplus z < 1$. Then $x \odot z = 0$, and (1.16) becomes $(x \odot y) \oplus z = (x \oplus z) \odot y$, which follows from Lemma 1.6.1(vii).

Subcase 3.2: $x \oplus y = x \oplus z = 1$.

Then equation (1.16) becomes

$$(1.19)\quad (x \odot y) \oplus z = (x \odot z) \oplus y.$$

Note that equation (1.19) holds in case $x \odot y = 0$ or $x \odot z = 0$. Indeed, suppose $x \odot y = 0$. Since $x \oplus y = 1$, it follows from Lemma 1.1.3 that $x = \neg y$, whence from $y = \neg x \leq z$ we obtain $(x \odot y) \oplus z = z = y \vee z =$

$(\neg y \odot z) \oplus y = (x \odot z) \oplus y$. By symmetry, (1.19) holds under the hypothesis $x \odot z = 0$. We next observe that if one of the members of (1.19) is equal to 1 then so is the other. Assume, for instance, $(x \odot y) \oplus z = 1$. Since $\neg x \oplus \neg y \oplus \neg z = 1$ is equivalent to $x \odot y \odot z = 0$, it follows from Lemma 1.1.3 that $z = \neg(x \odot y) = \neg x \oplus \neg y$. Hence, by Proposition 1.1.6, $(x \odot z) \oplus y = (x \odot (\neg x \oplus \neg y)) \oplus y = (x \wedge \neg y) \oplus y = (x \oplus y) \wedge (\neg y \oplus y) = 1$. Thus, to complete the analysis of Subcase 3.2 we may restrict to the case when $(x \odot y) \oplus z < 1$, $(x \odot z) \oplus y < 1$, $x \odot y > 0$, $x \odot z > 0$. Under these hypotheses, by Lemma 1.6.1(vii) we obtain:

$$x \odot (z \oplus (x \odot y)) = (x \odot z) \oplus (x \odot y) > 0,$$

$$x \odot (y \oplus (x \odot z)) = (x \odot y) \oplus (x \odot z),$$

thus establishing (1.19) by Lemma 1.6.1(iv).

Subcase 3.3: $x \oplus y < 1$ and $x \oplus z < 1$.

Then by Lemma 1.6.1(i), $x \odot y = 0$ and $x \odot z = 0$, i.e., $\neg x \oplus \neg y = 1$ and $\neg x \oplus \neg z = 1$. Recalling (1.18) and arguing as in Subcase 3.2 (with $\neg x, \neg y, \neg z$ instead of x, y, z) we conclude that (1.16) also holds in this case. □

1.7 Bibliographical remarks

MV-algebras were originally introduced by Chang in [36] to prove the completeness theorem for the Łukasiewicz calculus. In the same paper Chang proved the basic facts about the natural order, congruences and boolean elements. Prime ideals were introduced by him in [38] to prove Theorem 1.3.3. For historical information on MV-algebras see [40]. A version of Proposition 1.2.2 is in [86]. Equation (1.16) was proved in [51].

It is easy to see that the present definition of MV-algebras (which is essentially due to Mangani [144]) is equivalent to Chang's original definition. To this purpose one simply notes that for every MV-algebra $\langle A, \oplus, \neg, 0\rangle$ the operations $\vee$ and $\wedge$ defined by (1.5) and (1.6) are the join and the meet with respect to the natural order of A; thus in particular, both operations are commutative and associative.

(Equivalents of) MV-algebras are known in the literature under several names. As an example, following [249],[121], let us say that a *bounded commutative BCK-algebra* is an algebra $(A, *, 0, 1)$ with a binary (bounded subtraction) operation $*$ and two constants 0 and 1 satisfying the following equations:

Y1) $(x * y) * z = (x * z) * y$

Y2) $x * (x * y) = y * (y * x)$

Y3) $x * x = 0$

Y4) $x * 0 = x$

Y5) $x * 1 = 0.$

Bounded commutative BCK-algebras have been considered by several authors. See, e.g., [233], [213], [214], [116], [120].

A tedious but straightforward verification yields the following result, first proved in [86] (also see [164]):

Theorem 1.7.1 *If $(A, \oplus, \neg, 0)$ is an MV-algebra then $(A, \ominus, 0, 1)$ is a bounded commutative BCK-algebra. Moreover, we have the identities $\neg x = 1 \ominus x$ and $x \oplus y = 1 \ominus ((1 \ominus x) \ominus y))$.*

Conversely, for any bounded commutative BCK-algebra $(A, *, 0, 1)$, upon defining $\neg x =_{def} 1 * x$ and $x \oplus y =_{def} 1 * ((1 * x) * y)$, then $(A, \oplus, \neg, 0)$ is an MV-algebra, and $x \ominus y = x * y$.

There exist several other equivalent counterparts of MV-algebras, including Bosbach's *bricks* [31], Buff's *S-algebras* [33], Komori's *CN-algebras* [130], Lacava's *Ł-algebras* [131], Rodríguez's *Wajsberg algebras* [212], [86]. We will return to the latter in subsequent chapters. As we shall also see, up to categorical equivalence, MV-algebras are the same as abelian lattice-ordered groups with a distinguished strong unit. Thus MV-algebras provide an equational formulation of the theory of magnitudes with an archimedean unit.

Among all associative commutative continuous operations on the unit interval $[0, 1]$, the $\oplus$ operation and its dual $\odot$ have a special status. For details we refer, e.g., to the papers by Menu and Pavelka [153], [201], and to the relevant chapters of the book by Butnariu and Klement [34].

Chapter 2

Chang completeness theorem

In this chapter we shall prove Chang's completeness theorem stating that if an equation holds in the unit real interval $[0,1]$, then the equation holds in every MV-algebra. Thus, intuitively, the two element structure $\{0,1\}$ stands to boolean algebras as the interval $[0,1]$ stands to MV-algebras. Our proof is elementary, and makes use of tools (such as "good sequences") that shall also find applications in a subsequent chapter to show the equivalence between MV-algebras and lattice-ordered abelian groups with strong unit.

2.1 The functor Γ

A *partially ordered abelian group* is an abelian group $\langle G,+,-,0\rangle$ endowed with a partial order relation $\leq$ that is *compatible* with addition; in other words, $\leq$ has the following *translation invariance* property, for all $x,y,t\in G$:

(2.1) If $x\leq y$ then $t+x\leq t+y$.

The *positive cone* G^+ of G is the set of all $x\in G$ such that $0\leq x$. When the order relation is total, (i.e., when $G=G^+\cup -G^+$), G is said to be a *totally ordered abelian group*, or *o-group* for short. When the order of G defines a lattice structure, G is called a *lattice-ordered abelian group*, or *ℓ-group*, for short. In any ℓ-group we have

(2.2) $t+(x\vee y)=(t+x)\vee(t+y)$

and

(2.3) $t + (x \wedge y) = (t + x) \wedge (t + y)$.

In every o-group $x \vee y = \max\{x, y\}$ and $x \wedge y = \min\{x, y\}$; o-groups are particular cases of ℓ-groups.

For each element x of an ℓ-group G, the *positive part* x^+, the *negative part* x^-, and the *absolute value* $|x|$ of x are defined as follows:

(2.4) $x^+ =_{def} 0 \vee x$;

(2.5) $x^- =_{def} 0 \vee -x$;

(2.6) $|x| =_{def} x^+ + x^- = x \vee -x$.

A *strong (order) unit* u of G is an archimedean element of G, i.e., an element $0 \leq u \in G$ such that for each $x \in G$ there is an integer $n \geq 0$ with $|x| \leq nu$.

Definition 2.1.1 Let G be an ℓ-group. For any element $u \in G$, $u > 0$, (not necessarily u being a strong unit of G) we let

$$[0, u] =_{def} \{x \in G \mid 0 \leq x \leq u\},$$

and for each $x, y \in [0, u]$,

$$x \oplus y =_{def} u \wedge (x + y), \quad \text{and} \quad \neg x =_{def} u - x.$$

The structure $\langle [0, u], \oplus, \neg, 0 \rangle$ is denoted $\mathbf{\Gamma}(G, u)$.

Proposition 2.1.2 *$\mathbf{\Gamma}(G, u)$ is an MV-algebra.*

Proof: We shall limit ourselves to verifying that $\mathbf{\Gamma}(G, u)$ satisfies MV6). For all $x, y \in [0, u]$ we have

$$\neg(\neg x \oplus y) \oplus y$$

$$= y \oplus \neg(y \oplus \neg x)$$

$$= u \wedge (y + (u - (u \wedge (y + u - x))))$$

$$= u \wedge (y + u + (-u \vee (-y - u + x)))$$

$$= u \wedge ((y + u - u) \vee (y + u - y - u + x))$$

$$= u \wedge (y \vee x)$$

$$= y \vee x = x \vee y.$$

This shows that x and y are interchangeable. □

Lemma 2.1.3 *Let G be an ℓ-group with strong unit u. Let $A = \Gamma(G, u)$.*

(i) *For all $a, b \in A$, $a + b = (a \oplus b) + (a \odot b)$;*

(ii) *For all $x_1, \ldots, x_n \in A$, $x_1 \oplus \cdots \oplus x_n = u \wedge (x_1 + \cdots + x_n)$;*

(iii) *The natural order of the MV-algebra A coincides with the order of $[0, u]$ inherited from G by restriction.*

Proof: (i) We easily obtain $a+b-(a \odot b) = a+b-(0 \vee (a+b-u)) = (a+b) \wedge u = a \oplus b$. An easy induction on n proves (ii). (iii) is proved by direct verification, using 1.1.5, together with the above proof of 2.1.2. □

Notation: Following common usage, we let $\mathbf{R}, \mathbf{Q}, \mathbf{Z}$ denote the additive groups of reals, rationals, integers, with the natural order. In the particular case when $G = \mathbf{R}$, $\Gamma(\mathbf{R}, 1)$ coincides with the MV-algebra [0,1]. We also have $\mathbf{Q} \cap [0, 1] = \Gamma(\mathbf{Q}, 1)$. Further, for each integer $n \geq 2$,

$$Ł_n = \Gamma(\mathbf{Z}\frac{1}{n-1}, 1), \text{ where } \mathbf{Z}\frac{1}{n-1} = \{\frac{z}{n-1} \mid z \in \mathbf{Z}\}.$$

In particular, the boolean algebra $\{0, 1\} = Ł_2$ is the same as $\Gamma(\mathbf{Z}, 1)$.

When dealing with elements $x \in \Gamma(G, u)$, the notation nx may ambiguously represent both $x + \cdots + x$ (n times), and $x \oplus \cdots \oplus x$ (n times). To avoid any danger of confusion, we will adopt the following notation:

$$(2.7) \quad n.x =_{def} \underbrace{x \oplus \cdots \oplus x}_{n \text{ times}} = u \wedge \underbrace{(x + \cdots + x)}_{n \text{ times}} = u \wedge nx.$$

Let G be an ℓ-group and $0 < u \in G$. Let $S = \{x \in G \mid \text{for some } 0 \leq n \in \mathbf{Z}, |x| \leq nu\}$. Then S is a subgroup and a sublattice of G containing u, and $\mathbf{\Gamma}(G, u) = \mathbf{\Gamma}(S, u)$. Therefore, when considering the MV-algebras $\mathbf{\Gamma}(G, u)$ we can safely assume that u is a strong unit of G.

Definition 2.1.4 Let G and H be ℓ-groups. A function $h: G \to H$ is said to be an *ℓ-group homomorphism* iff h is both a group-homomorphism and a lattice-homomorphism; in other words, for each $x, y \in G$, $h(x-y) = h(x)-h(y)$, $h(x \vee y) = h(x) \vee h(y)$ and $h(x \wedge y) = h(x) \wedge h(y)$. Suppose that $0 < u \in G$ and $0 < v \in H$, and let $h: G \to H$ be an ℓ-group homomorphism such that $h(u) = v$. Then h is said to be a *unital* ℓ-homomorphism.

Letting $\mathbf{\Gamma}(h) =_{def} h|_{[0,u]}$ denote the restriction of h to the unit interval $[0, u]$, then $\mathbf{\Gamma}(h)$ is a homomorphism from $\mathbf{\Gamma}(G, u)$ into $\mathbf{\Gamma}(H, v)$.

Proposition 2.1.5 *Let $\mathcal{A}$ denote the category whose objects are pairs $\langle G, u \rangle$ with G an ℓ-group and u a distinguished strong unit of G, and whose morphisms are unital ℓ-homomorphisms. Then $\mathbf{\Gamma}$ is a functor from $\mathcal{A}$ into the category $\mathcal{MV}$ of MV-algebras.* □

The above result shall be strengthened in a subsequent chapter, where we shall prove that $\mathbf{\Gamma}$ is a natural equivalence (i.e., a full, faithful, dense functor) between $\mathcal{A}$ and $\mathcal{MV}$.

2.2 Good sequences

A sequence $\mathbf{a} = (a_1, a_2, \ldots)$ of elements of an arbitrary MV-algebra A is said to be *good* iff for each $i = 1, 2 \ldots$,

$$a_i \oplus a_{i+1} = a_i,$$

and there is an integer n such that $a_r = 0$ for all $r > n$. Instead of $\mathbf{a} = (a_1, \ldots, a_n, 0, 0, \ldots)$ we shall often write, more concisely,

$$\mathbf{a} = (a_1, \ldots, a_n).$$

Thus, if 0^m denotes an m-tuple of zeros, we have identical good sequences

(2.8) $(a_1, \ldots, a_n) = (a_1, \ldots, a_n, 0^m)$.

Note that by *pre*-pending to the good sequence $(a_1, \ldots, a_n)$ an m-tuple 1^m of consecutive ones, the resulting good sequence $(1^m, a_1, \ldots, a_n)$ is different from $(a_1, \ldots, a_n)$. For each $a \in A$, the good sequence $(a, 0 \ldots, 0, \ldots)$ will be denoted by (a).

The good sequences of a boolean algebra A are the nonincreasing sequences of elements of A having a finite number of nonzero terms.

For totally ordered MV-algebras we have the following characterization of good sequences:

Proposition 2.2.1 *If A is an MV-chain then each good sequence of A has the form*

(2.9) $(1^p, a)$ *for some integer* $p \geq 0$ *and* $a \in A$.

Proof: Immediate from Lemma 1.6.1(v). □

Lemma 2.2.2 *Suppose that $A \subseteq \prod_i A_i$ is the subdirect product of a family $\{A_i\}_{i \in I}$ of MV-algebras. A sequence $\mathbf{a} = (a_1, \ldots, a_n, \ldots)$ of elements of A is a good sequence if and only if for each $i \in I$ the sequence*

$$\mathbf{a}_i = (\pi_i(a_1), \ldots, \pi_i(a_n), \ldots)$$

is a good sequence in A_i, and there is an integer $n_0 \geq 0$ such that whenever $n > n_0$ then for all $i \in I$, $\pi_i(a_n) = 0$.

Proof: It is sufficient to note that $a_n \oplus a_{n+1} = a_n$ means that $\pi_i(a_n \oplus a_{n+1}) = \pi_i(a_n)$ for each $i \in I$. □

In the light of Theorem 1.3.3, the above Lemma 2.2.2 and Proposition 2.2.1 yield a very useful tool for dealing with good sequences. As an example, consider the proof of the following result:

Lemma 2.2.3 *Let A be an MV-algebra. If $\mathbf{a} = (a_1, \ldots, a_n, \ldots)$ and $\mathbf{b} = (b_1, \ldots, b_n, \ldots)$ are good sequences of A, then so is the sequence $\mathbf{c} = (c_1, \ldots, c_n, \ldots)$ given by $c_n = a_n \vee b_n$ for each n.*

Proof: Since $\mathbf{a}$ and $\mathbf{b}$ are both good sequences, there is an integer n_0 such that $c_n = 0$ for all $n > n_0$. By Theorem 1.3.3, A is a subdirect product of a family $\{C_i\}_{i\in I}$ of MV-chains. For each $i \in I$ the sequences $\mathbf{a}_i = (\pi_i(a_1), \ldots, \pi_i(a_n), \ldots)$ and $\mathbf{b}_i = (\pi_i(b_1), \ldots, \pi_i(b_n), \ldots)$ are good sequences of C_i. Hence, by Proposition 2.2.1, $\mathbf{a}_i = (1^p, \alpha_i)$ and $\mathbf{b}_i = (1^q, \beta_i)$, where α_i and β_i are in C_i. Therefore, $\pi_i(c_n) = 1$ if $n \leq \max\{p,q\}$ and $\pi_i(c_n) = 0$ if $n > \max\{p,q\} + 1$. For $n = \max\{p,q\} + 1$, we have $\pi_i(c_n) = \alpha_i$ if $p > q$, $\pi_i(c_n) = \beta_i$ if $p < q$ and $\pi_i(c_n) = \max\{\alpha_i, \beta_i\}$ when $p = q$. Consequently, letting $\mathbf{c}_i = (\pi_i(c_1), \ldots, \pi_i(c_n), \ldots)$ it follows that $\mathbf{c}_i$ is a good sequence for each $i \in I$, whence we conclude that $\mathbf{c}$ is a good sequence of A. □

Example: To have a better intuition of the meaning of good sequences, for every real number $\alpha \geq 0$ let $\lfloor\alpha\rfloor$ denote the greatest integer $\leq \alpha$, and let $\langle\alpha\rangle =_{def} \alpha - \lfloor\alpha\rfloor$ be the fractional part of α. There will be no danger of confusion between this notation and the notation for principal ideals. Then α can be written as

$$\alpha = 1 + \ldots + 1 + \langle\alpha\rangle + 0 + 0 + \ldots$$

with $\lfloor\alpha\rfloor$ many consecutive 1's. Considered as elements of the MV-algebra $[0,1]$, the above summands $\alpha_1, \alpha_2, \ldots$ of α trivially satisfy the identity $\alpha_i \oplus \alpha_{i+1} = \alpha_i$ for every integer $i \geq 1$. For $0 \leq \beta \in \mathbf{R}$, let similarly

$$\beta = \beta_1 + \ldots + \beta_{m-1} + \langle\beta\rangle + 0 + \ldots,$$

where $\beta_1 = \ldots = \beta_{m-1} = 1 = \alpha_1 = \ldots = \alpha_{n-1}$, $0 = \alpha_{n+1} = \alpha_{n+2} = \ldots$, and $0 = \beta_{m+1} = \beta_{m+2} = \ldots$. Let $\gamma = \alpha + \beta$. Then $\gamma = \gamma_1 + \gamma_2 + \ldots$, where $\gamma_1 = \ldots = \gamma_{n+m-2} = 1$, $\gamma_{n+m-1} = \langle\alpha\rangle \oplus \langle\beta\rangle$, $\gamma_{n+m} = \langle\alpha\rangle \odot \langle\beta\rangle$, and $0 = \gamma_{n+m+1} = \gamma_{n+m+2} = \ldots$. In a more compact notation, for each $i = 1, 2, \ldots$, the summand γ_i is given by

(2.10) $\gamma_i = \alpha_i \oplus (\alpha_{i-1} \odot \beta_1) \oplus (\alpha_{i-2} \odot \beta_2) \oplus \ldots \oplus (\alpha_2 \odot \beta_{i-2}) \oplus (\alpha_1 \odot \beta_{i-1}) \oplus \beta_i$.

In the light of (2.10) and (2.8), we can now give the following

Definition 2.2.4 For any two good sequences $\mathbf{a} = (a_1, \ldots, a_n)$ and $\mathbf{b} = (b_1, \ldots, b_m)$ their *sum* $\mathbf{c} = \mathbf{a} + \mathbf{b}$ is defined by $\mathbf{c} = (c_1, c_2, \ldots)$, where for all $i = 1, 2, \ldots$

(2.11) $c_i =_{def} a_i \oplus (a_{i-1} \odot b_1) \oplus \ldots \oplus (a_1 \odot b_{i-1}) \oplus b_i$.

Since $a_p = b_q = 0$ whenever $p > n$ and $q > m$, then c_j identically vanishes for each $j > m+n$. The notation $\mathbf{c} = (c_1, \ldots, c_{n+m}) = (a_1, \ldots, a_n) + (b_1, \ldots, b_m)$ is self-explanatory.

The following immediate consequence of (2.11) will be frequently used to compute the sum of two good sequences in an MV-chain:

$$(2.12) \quad (1^p, a) + (1^q, b) = (1^{p+q},\ a \oplus b,\ a \odot b).$$

2.3 The partially ordered monoid M_A

Since by equation (1.14), $(a \oplus b, a \odot b)$ is a good sequence, applying Theorem 1.3.3 and Lemma 2.2.2 together with (2.12), we immediately get that *the sum of two good sequences is a good sequence.* We denote by M_A the set of good sequences of A equipped with addition.

Proposition 2.3.1 *Let A be an MV-algebra. Then M_A is an abelian monoid with the following additional properties:*

(i) (cancellation) *For any good sequences* $\mathbf{a}$, $\mathbf{b}$, $\mathbf{c}$, *if* $\mathbf{a}+\mathbf{b} = \mathbf{a}+\mathbf{c}$ *then* $\mathbf{b} = \mathbf{c}$*;*

(ii) (zero-law) *If* $\mathbf{a}+\mathbf{b} = (0)$ *then* $\mathbf{a} = \mathbf{b} = (0)$.

Proof: By (2.11), $\mathbf{a}+(0) = \mathbf{a}$, addition is commutative, and the zero-law holds. To prove associativity, by Theorem 1.3.3 we can safely assume A to be totally ordered. By Proposition 2.2.1 and equation (1.16) in Proposition 1.6.2, letting $\mathbf{a} = (1^p, a)$, $\mathbf{b} = (1^q, b)$, and $\mathbf{c} = (1^r, c)$, we have the identities

$$\begin{aligned}
&(\mathbf{b}+\mathbf{a})+\mathbf{c} \\
&= (1^{p+q+r},\ a \oplus b \oplus c,\ (a \odot b) \oplus ((a \oplus b) \odot c),\ a \odot b \odot c) \\
&= (1^{p+q+r},\ a \oplus b \oplus c,\ (a \odot c) \oplus ((a \oplus c) \odot b),\ a \odot b \odot c) \\
&= \mathbf{b}+(\mathbf{a}+\mathbf{c}).
\end{aligned}$$

Similarly, to prove cancellation, avoiding trivialities, assume that a, b and c are different from 1. If $q = r$, then by Lemma 1.6.1(ii), $b = c$, and we are done. If $q < r-1$ then from the identity $(1^{p+q}, a\oplus b, a\odot b) = (1^{p+r}, a \oplus c, a \odot c)$ we get $a \odot b = 1$, i.e., $a = b = 1$, which is a contradiction. If $q = r-1$ then $a \odot b = a$ and $a \oplus b = 1$, which is impossible because these two equalities imply that $b = 1$. The cases corresponding to $r < q$ are similarly shown to lead to contradiction. □

Proposition 2.3.2 *Let* $\mathbf{a} = (a_1, \ldots, a_n)$ *and* $\mathbf{b} = (b_1, \ldots, b_m)$ *be good sequences. Recalling* (2.8) *assume, without loss of generality,* $m = n$. *Then the following are equivalent:*

(i) *There is a good sequence* $\mathbf{c}$ *such that* $\mathbf{b} + \mathbf{c} = \mathbf{a}$;

(ii) $b_i \leq a_i$ *for all* $i = 1, \ldots, n$.

Proof: (i) ⇒ (ii) is immediate from (2.11). (ii) ⇒ (i). Observe that by the remark following Lemma 1.6.1, $(\neg b_n, \ldots, \neg b_1)$ is a good sequence. Let us denote by $\mathbf{c} = \mathbf{a} - \mathbf{b}$ the good sequence obtained by dropping the first n terms in $(a_1, \ldots, a_n) + (\neg b_n, \ldots, \neg b_1)$. We shall prove that $\mathbf{c} + \mathbf{b} = \mathbf{a}$. Using Theorem 1.3.3 we can safely assume A to be totally ordered, so that by (2.9), $\mathbf{a} = (1^p, a)$ and $\mathbf{b} = (1^q, b)$. To avoid trivialities assume both a and b to be different from 0 and from 1. Then $q \leq p$. Upon rewriting $\mathbf{b} = (1^q,\ b,\ 0^{p-q})$, from $n = p+1$ we get $(\neg b_n, \ldots, \neg b_1) = (1^{p-q},\ \neg b,\ 0^q)$, and hence $\mathbf{c}$ is obtained by dropping the first $p+1$ terms from $(1^{2p-q},\ a \oplus \neg b,\ a \ominus b)$.

Case 1: $b \leq a$.

Then $a \oplus \neg b = 1$, $\mathbf{c} = (1^{p-q}, a \ominus b)$ and $\mathbf{c} + \mathbf{b} = (1^p,\ (a \ominus b) \oplus b,\ (a \ominus b) \odot b) = (1^p,\ b \vee a,\ 0) = (1^p,\ a) = \mathbf{a}$.

Case 2: $b > a$.

Then $p > q$, $a \ominus b = 0$, $\mathbf{c} = (1^{p-q-1},\ a \oplus \neg b)$ and $\mathbf{c} + \mathbf{b} = (1^{p-1},\ a \oplus \neg b \oplus b, (a \oplus \neg b) \odot b) = (1^p, a \wedge b) = (1^p, a) = \mathbf{a}$. □

Definition 2.3.3 Given any two good sequences $\mathbf{a}$ and $\mathbf{b}$ of A we write:

(2.13) $\mathbf{b} \leq \mathbf{a}$ iff $\mathbf{b}$ and $\mathbf{a}$ satisfy the equivalent conditions of 2.3.2.

Proposition 2.3.4 *Let* $\mathbf{a}$ *and* $\mathbf{b}$ *be good sequences.*
(i) If $\mathbf{b} \leq \mathbf{a}$ *then there is a unique good sequence* $\mathbf{c}$ *such that* $\mathbf{b}+\mathbf{c} = \mathbf{a}$. *This* $\mathbf{c}$, *denoted* $\mathbf{a} - \mathbf{b}$, *is given by*

(2.14) $\mathbf{c} = (a_1, \ldots, a_n) + (\neg b_n, \ldots, \neg b_1)$ *omitting the first n terms.*

(ii) In particular, for each $a \in A$ *we have*

(2.15) $(\neg a) = (1) - (a)$.

(iii) The order is translation invariant, *in the sense that* $\mathbf{b} \leq \mathbf{a}$ *implies* $\mathbf{b} + \mathbf{d} \leq \mathbf{a} + \mathbf{d}$ *for every good sequence* $\mathbf{d}$.

Proof: By an easy adaptation of the proof of Proposition 2.3.2, together with Proposition 2.3.1 (i). □

Proposition 2.3.5 *Let* $\mathbf{a} = (a_1, \ldots, a_n, \ldots)$ *and* $\mathbf{b} = (b_1, \ldots, b_n, \ldots)$ *be good sequences of an MV-algebra A.*
(i) The sequence

$$\mathbf{a} \vee \mathbf{b} =_{def} (a_1 \vee b_1, \ldots, a_n \vee b_n, \ldots)$$

is good, and is in fact the supremum of $\mathbf{a}$ *and* $\mathbf{b}$ *with respect to the order defined by (2.13).*
(ii) Analogously, the good sequence

$$\mathbf{a} \wedge \mathbf{b} =_{def} (a_1 \wedge b_1, \ldots, a_n \wedge b_n, \ldots)$$

is the infimum of $\mathbf{a}$ *and* $\mathbf{b}$.
(iii) For all $a, b, c \in A$ *we have*

(2.16) $((a) + (b)) \wedge (1) = (a \oplus b)$.

Proof: By Lemma 2.2.3, together with Proposition 2.3.2(ii) and (2.11). □

2.4 Chang's ℓ-group G_A

From the abelian monoid M_A, enriched with the lattice-order of Proposition 2.3.5, one can routinely obtain an ℓ-group G_A such that M_A is isomorphic, both as a monoid and as a lattice, to the positive cone ${G_A}^+$. To this purpose, mimicking the construction of $\mathbf{Z}$ from $\mathbf{N}$, let us agree to say that a pair of good sequences $(\mathbf{a}, \mathbf{b})$ is *equivalent* to another pair $(\mathbf{a}', \mathbf{b}')$ iff $\mathbf{a} + \mathbf{b}' = \mathbf{a}' + \mathbf{b}$. Transitivity of this relation follows from cancellation, Proposition 2.3.1(i).

Notation: The equivalence class of the pair $(\mathbf{a}, \mathbf{b})$ shall be denoted by

$$[\mathbf{a}, \mathbf{b}].$$

There will be no danger of confusion with the notation for unit intervals in ℓ-groups.

Let $G_A = \langle G_A, 0, +, -\rangle$ be the set of equivalence classes of pairs of good sequences, where

the *zero element* 0 is the equivalence class $[(0), (0)]$,

addition is defined by $[\mathbf{a}, \mathbf{b}] + [\mathbf{c}, \mathbf{d}] =_{def} [\mathbf{a} + \mathbf{c}, \mathbf{b} + \mathbf{d}]$, and

subtraction is defined by $-[\mathbf{a}, \mathbf{b}] =_{def} [\mathbf{b}, \mathbf{a}]$.

Then by direct inspection one easily sees that G_A is an abelian group. G_A is called the *enveloping group* of A.

We shall now equip G_A with a lattice-order. Let $(\mathbf{a}, \mathbf{b})$ be a pair of good sequences of the MV-algebra A. By Proposition 2.3.2(i), $(\mathbf{a}, \mathbf{b})$ has an equivalent pair of the form $(\mathbf{e}, (0))$ if and only if $\mathbf{a} \geq \mathbf{b}$. Let M'_A be the submonoid of G_A given by the equivalence classes of pairs $(\mathbf{e}, (0))$, for all good sequences $\mathbf{e}$. Since the map $\mathbf{e} \mapsto (\mathbf{e}, (0))$ induces an isomorphism of the monoid M_A onto M'_A, we shall freely identify the two monoids M_A and M'_A.

Definition 2.4.1 Let A be an MV-algebra, and $\mathbf{a}, \mathbf{b}, \mathbf{c}, \mathbf{d} \in M_A$. We say that the equivalence class $[\mathbf{c}, \mathbf{d}]$ *dominates* the equivalence class $[\mathbf{a}, \mathbf{b}]$, in symbols,

$$[\mathbf{a}, \mathbf{b}] \preceq [\mathbf{c}, \mathbf{d}],$$

iff $[\mathbf{c}, \mathbf{d}]-[\mathbf{a}, \mathbf{b}] = [\mathbf{e}, (0)]$ for some good sequence $\mathbf{e} \in M_A$. Equivalently, $[\mathbf{a}, \mathbf{b}] \preceq [\mathbf{c}, \mathbf{d}]$ iff $\mathbf{a}+\mathbf{d} \leq \mathbf{c}+\mathbf{b}$, where $\leq$ is the partial order of M_A given by Definition 2.3.3.

Proposition 2.4.2 *Let A be an MV-algebra.*
(i) The relation $\preceq$ is a translation invariant partial order, making G_A into an ℓ-group. Specifically, for any two pairs of good sequences $(\mathbf{a}, \mathbf{b})$ and $(\mathbf{c}, \mathbf{d})$ the supremum of their equivalence classes in G_A is the equivalence class of $((\mathbf{a}+\mathbf{d}) \vee (\mathbf{c}+\mathbf{b}), \mathbf{b}+\mathbf{d})$, where $\vee$ is the supremum in M_A given by Proposition 2.3.5. In symbols,

$$(2.17)\quad [\mathbf{a}, \mathbf{b}] \bigvee [\mathbf{c}, \mathbf{d}] = [(\mathbf{a}+\mathbf{d}) \vee (\mathbf{c}+\mathbf{b}), \mathbf{b}+\mathbf{d}].$$

(ii) Similarly, the infimum $[\mathbf{a}, \mathbf{b}] \bigwedge [\mathbf{c}, \mathbf{d}]$ is given by

$$(2.18)\quad [\mathbf{a}, \mathbf{b}] \bigwedge [\mathbf{c}, \mathbf{d}] = [(\mathbf{a}+\mathbf{d}) \wedge (\mathbf{c}+\mathbf{b}), \mathbf{b}+\mathbf{d}].$$

(iii) The map $\mathbf{a} \in M_A \mapsto [\mathbf{a}, (0)]$ is an isomorphism between the monoid M_A, equipped with the lattice-order of Proposition 2.3.5, and the positive cone ${G_A}^{+} =_{def} \{[\mathbf{c}, \mathbf{d}] \in G_A \mid \mathbf{c} \geq \mathbf{d}\}$, with the lattice-order inherited by restriction of $\preceq$.

Proof: (i) The proof that $\preceq$ is a translation invariant partial order on G_A is routine. In order to prove (2.17), first of all, from the inequality $\mathbf{a}+\mathbf{d} \leq (\mathbf{a}+\mathbf{d}) \vee (\mathbf{c}+\mathbf{b})$ we obtain $[\mathbf{a}, \mathbf{b}] \preceq [(\mathbf{a}+\mathbf{d}) \vee (\mathbf{c}+\mathbf{b}), \mathbf{b}+\mathbf{d}]$, and, symmetrically, $[\mathbf{c}, \mathbf{d}] \preceq [(\mathbf{a}+\mathbf{d}) \vee (\mathbf{c}+\mathbf{b}), \mathbf{b}+\mathbf{d}]$. Thus, $[(\mathbf{a}+\mathbf{d}) \vee (\mathbf{c}+\mathbf{b}), \mathbf{b}+\mathbf{d}]$ is an upper bound of $[\mathbf{a}, \mathbf{b}]$ and $[\mathbf{c}, \mathbf{d}]$. To show that this is indeed *the least* upper bound, for any upper bound $[\mathbf{p}, \mathbf{q}]$ we must find an element $\mathbf{z} \in M_A$ such that

$$(2.19)\quad \mathbf{p}+\mathbf{d}+\mathbf{b} = \mathbf{z}+\mathbf{q}+((\mathbf{a}+\mathbf{d}) \vee (\mathbf{b}+\mathbf{c})).$$

By hypothesis, there are $\mathbf{x}, \mathbf{y} \in M_A$ such that $\mathbf{p}+\mathbf{b} = \mathbf{x}+\mathbf{q}+\mathbf{a}$ and $\mathbf{p}+\mathbf{d} = \mathbf{y}+\mathbf{q}+\mathbf{c}$. Let $\mathbf{z} \in M_A$ be such that $\mathbf{x}+\mathbf{y} = \mathbf{z}+(\mathbf{x} \vee \mathbf{y})$; the existence of $\mathbf{z}$ is ensured by the inequality $\mathbf{x}+\mathbf{y} \geq \mathbf{x} \vee \mathbf{y}$, using Propositions 2.3.2 and 2.2.3. One now establishes (2.19) using the cancellation property of M_A, as follows:

$$2\mathbf{p}+\mathbf{b}+\mathbf{d}$$

$$= \mathbf{2q} + \mathbf{a} + \mathbf{c} + \mathbf{x} + \mathbf{y}$$
$$= \mathbf{2q} + \mathbf{a} + \mathbf{c} + \mathbf{z} + (\mathbf{x} \vee \mathbf{y})$$
$$= \mathbf{z} + \mathbf{q} + ((\mathbf{x} + \mathbf{q} + \mathbf{a} + \mathbf{c}) \vee (\mathbf{y} + \mathbf{q} + \mathbf{a} + \mathbf{c}))$$
$$= \mathbf{z} + \mathbf{q} + ((\mathbf{p} + \mathbf{b} + \mathbf{c}) \vee (\mathbf{p} + \mathbf{d} + \mathbf{a}))$$
$$= \mathbf{p} + \mathbf{z} + \mathbf{q} + ((\mathbf{b} + \mathbf{c}) \vee (\mathbf{d} + \mathbf{a})).$$

One similarly proves (ii). Finally, (iii) is an immediate consequence of the definitions of the partial orders $\leq$ and $\preceq$. □

Definition 2.4.3 The ℓ-group G_A with the above lattice-order is called the *Chang ℓ-group* of the MV-algebra A.

Proposition 2.4.4 *The element $u_A =_{def} [(1),(0)]$ is a strong unit of the ℓ-group G_A.*

Proof: As a matter of fact, any element of ${G_A}^+$ can be represented by $[\mathbf{a},(0)]$, for some good sequence $\mathbf{a} = (a_1, a_2, \ldots)$ in A. Let the integer $m \geq 1$ be so chosen that $a_n = 0$ for all $n \geq m$. By Definition 2.4.1, $mu_A = [1^m, (0)]$ dominates $[\mathbf{a},(0)]$, whence the desired conclusion immediately follows. □

A crucial property of the ℓ-group G_A is given by the following result:

Theorem 2.4.5 *The correspondence*

$$a \mapsto \varphi_A(a) = [(a),(0)]$$

defines an isomorphism from the MV-algebra A onto the MV-algebra $\Gamma(G_A, u_A)$.

Proof: By definition, $[(0),(0)] \preceq [\mathbf{a},\mathbf{b}] \preceq u_A$ iff there is $c \in A$ such that $(\mathbf{a},\mathbf{b})$ is equivalent to $((c),(0))$. Thus, φ_A maps A onto the unit interval $[[(0),(0)],u_A]$ of G_A. It is easy to see that this map is one-one. By (2.16), $\varphi_A(a \oplus b) = (\varphi_A(a) + \varphi_A(b)) \wedge u_A$, and by (2.15), $\varphi_A(\neg a) = u_A - \varphi_A(a)$. Therefore, φ_A is a homomorphism from A to $\Gamma(G_A, u_A)$. □

Remark: An MV-algebra A is a chain if and only if G_A is totally ordered. Indeed, if A is totally ordered, then it follows from Proposition 2.3.2(i) that M_A is totally ordered, and this implies that G_A is a totally ordered group. The converse is an immediate consequence of the Theorem 2.4.5 above.

2.5 Chang completeness theorem

An *ℓ-group term* in the variables $x_1, \ldots, x_t$ is a string of symbols over the alphabet $\{x_1, \ldots, x_n, 0, -, +, \vee, \wedge, (,)\}$ which is obtained by the same inductive procedure used in Chapter 1.4 to define MV-terms.

Let τ be an ℓ-group term in the variables $x_1, \ldots, x_t$ and G be an ℓ-group. Substituting an element $a_i \in G$ for all occurrences of the variable x_i in τ, for $i = 1, \ldots, t$, and interpreting the symbols 0, $-$, $+$, $\vee$ and $\wedge$ as the corresponding operations in G, we obtain an element of G, denoted $\tau^G(a_1, \ldots, a_t)$.

To each MV-term τ in the n variables $x_1, \ldots, x_n$ we associate an ℓ-group term $\hat{\tau}$ in the $n+1$ variables $(x_1, \ldots, x_n, y)$, according to the following stipulations:

$$\hat{x}_i =_{def} x_i, \text{ for each } i = 1, \ldots, n,$$

$$\hat{0} =_{def} 0,$$

$$\widehat{\neg\sigma} =_{def} (y - \hat{\sigma}),$$

$$\widehat{(\rho \oplus \sigma)} =_{def} (y \wedge (\hat{\rho} + \hat{\sigma})).$$

Since unique readability also holds for ℓ-group terms, the mapping $\tau \mapsto \hat{\tau}$ is well defined; indeed, a moment's reflection shows that this map is computable by a Turing machine. We then have a purely syntactic counterpart of the mappings $(G, u) \mapsto \Gamma(G, u)$ and $A \mapsto G_A$, in a sense that is made precise by the following two propositions:

Proposition 2.5.1 *If G is a totally ordered abelian group, $0 < u \in G$, $0 \leq g_1, \ldots g_n \leq u$ and $A = \Gamma(G, u)$, then for every MV-term $\tau(x_1, \ldots, x_n)$ we have $\tau^A(g_1, \ldots, g_n) = \hat{\tau}^G(g_1, \ldots, g_n, u)$.*

Proof: By induction on the number of operation symbols in τ. The basis is trivial. For the induction step, by definition of Γ we have:

$$(\neg\sigma)^A(g_1, \ldots, g_n)$$

$$= \neg(\sigma^A(g_1, \ldots, g_n))$$

$$= \neg(\hat{\sigma}^G(g_1, \ldots, g_n, u))$$

$$= u - \hat{\sigma}^G(g_1, \ldots, g_n, u)$$

$$= (y^G - \hat{\sigma}^G)(g_1, \ldots, g_n, u)$$

$$= \widehat{\neg\sigma}^G(g_1, \ldots, g_n, u).$$

The $\oplus$-case is similar. □

In the light of Proposition 2.4.2(iii), we shall now identify M_A with the positive cone of G_A :

Proposition 2.5.2 *If A is an MV-chain, $a_1, \ldots, a_n$ are elements of A, and $\tau(x_1, \ldots, x_n)$ is an MV-term, then the one-term sequence $(\tau^A(a_1, \ldots, a_n)) \in M_A \subseteq G_A^+$ coincides with $\hat{\tau}^{G_A}((a_1), \ldots, (a_n), (1))$.*

Proof: By induction on the number of operation symbols in τ. The basis is trivial. For the induction step, if $\sigma = \phi \oplus \psi$ then using (2.16), together with the definition of the mapping $\sigma \mapsto \hat{\sigma}$, and omitting unnecessary superscripts, we can write:

$$(\phi \oplus \psi(a_1, \ldots, a_n)) = (\phi(a_1, \ldots, a_n) \oplus \psi(a_1, \ldots, a_n))$$

$$= \min((1), (\phi(a_1, \ldots, a_n)) + (\psi(a_1, \ldots, a_n)))$$

$$= \min((1), \hat{\phi}((a_1), \ldots, (a_n), (1)) + \hat{\psi}((a_1), \ldots, (a_n), (1)))$$

$$= (y \wedge (\hat{\phi} + \hat{\psi}))((a_1), \ldots, (a_n), (1))$$

$$= \widehat{\phi \oplus \psi}((a_1), \ldots, (a_n), (1)).$$

In the $\neg$-case, one similarly uses (2.15). □

Theorem 2.5.3 (Completeness Theorem) *An equation holds in $[0, 1]$ if and only if it holds in every MV-algebra.*

Proof: Suppose an equation fails in an MV-algebra A. By the remark following Definition 1.4.4 we may assume that the equation has the form $\tau(x_1,\ldots,x_n)=0$. By Corollary 1.4.7, A may be assumed to be totally ordered. There are elements $a_1,\ldots,a_n\in A$ such that $\tau^A(a_1,\ldots,a_n)>0$. Letting G_A denote the Chang ℓ-group of A, and again writing $M_A=G_A^+$, by Proposition 2.5.2 we have $0<\hat{\tau}^{G_A}((a_1),\ldots,(a_n),(1))\leq(1)$. Let $S=\mathbf{Z}(1)+\mathbf{Z}(a_1)+\cdots+\mathbf{Z}(a_n)$ be the subgroup of G_A generated by the elements $(1),(a_1),\ldots,(a_n)$, with the induced total order.

Since the order in G_A is translation invariant, it follows that G_A is torsion-free. Since S is a finitely generated subgroup of G_A, by the fundamental theorem on torsion-free abelian groups, we can identify S with the free abelian group $\mathbf{Z}^r$, for some integer $r\geq 1$. Its elements (1), $(a_1),\ldots,(a_n)$ are respectively identified with vectors $\mathbf{h}_0,\mathbf{h}_1,\ldots,\mathbf{h}_n\in\mathbf{Z}^r$; the set of nonnegative elements of S then becomes a submonoid P of $\mathbf{Z}^r$ such that

$$(2.20)\qquad P\cap -P=\{0\}\quad\text{and}\quad P\cup -P=\mathbf{Z}^r.$$

For any two vectors $\mathbf{h},\mathbf{k}\in\mathbf{Z}^r$ let us write

$$\mathbf{h}\leq_P\mathbf{k}\quad\text{iff}\quad \mathbf{k}-\mathbf{h}\in P.$$

Let us display the subterms $\sigma_0,\sigma_1,\ldots,\sigma_t$ of $\hat{\tau}$ as follows:

$$\sigma_0=y,\ \sigma_1=x_1,\ \ldots,\sigma_n=x_n,\ \sigma_{n+1},\ \sigma_{n+2},\ \ldots,\sigma_{t-1},\ \sigma_t=\hat{\tau}.$$

We can safely assume that the list contains the zero term. The map $y\mapsto\mathbf{h}_0$, $x_1\mapsto\mathbf{h}_1,\ldots,x_n\mapsto\mathbf{h}_n$ uniquely extends to an interpretation $\sigma_j\mapsto\mathbf{h}_j$ $(j=0,\ldots,t)$ of subterms of $\hat{\tau}$ into elements of the totally ordered group $T=(\mathbf{Z}^r,\leq_P)$. In particular, by hypothesis we have

$$(2.21)\qquad 0\leq_P\mathbf{h}_1,\ldots,\mathbf{h}_n\leq_P\mathbf{h}_0,\quad 0\leq_P\mathbf{h}_t\leq_P\mathbf{h}_0,$$

$$(2.22)\qquad 0\neq\mathbf{h}_t=\hat{\tau}^T(\mathbf{h}_1,\ldots,\mathbf{h}_n,\mathbf{h}_0).$$

Let ω be a permutation of $\{0,\ldots,t\}$ such that

$$(2.23)\qquad \mathbf{h}_{\omega(0)}\ \leq_P\ \mathbf{h}_{\omega(1)}\ \leq_P\ldots\leq_P\ \mathbf{h}_{\omega(t)}.$$

Our aim is to replace $\leq_P$ by another total order $\leq_{P'}$ over $\mathbf{Z}^r$ in such a way that the above inequalities still hold with respect to $\leq_{P'}$, and $(\mathbf{Z}^r, \leq_{P'})$ is isomorphic to a subgroup of the additive group $\mathbf{R}$ with the natural order. For each $j = 1, \ldots, t$, let the vector $\mathbf{d}_j \in P$ be defined by

$$\mathbf{d}_j = \mathbf{h}_{\omega(j)} - \mathbf{h}_{\omega(j-1)} = (d_{j1}, d_{j2}, \ldots, d_{jr}).$$

Embedding $\mathbf{Z}^r$ into $\mathbf{R}^r$, we define the positive and the negative span of the $\mathbf{d}_j$'s as follows:

$$(2.24) \qquad P^* = \{\sum_{j=1}^{t} \lambda_j \mathbf{d}_j \mid 0 \leq \lambda_j \in \mathbf{R}\}, \quad N^* = -P^*.$$

Note that P^* is a closed and convex subset of $\mathbf{R}^r$, and whenever $\mathbf{h} \in P^*$ and $0 \leq \alpha \in \mathbf{R}$, then $\alpha\mathbf{h} \in P^*$.

Claim 1: Whenever $\lambda_1, \ldots, \lambda_t$ are real numbers ≥ 0 and $\sum_{i=1}^{t} \lambda_i \mathbf{d}_i = 0$, it follows that $\lambda_i = 0$ for each i such that $\mathbf{d}_i \neq 0$.

Otherwise (absurdum hypothesis) let $I = \{i \in \{1, \ldots, t\} \mid \mathbf{d}_i \neq 0\}$, and assume $i^* \in I$, $\lambda_{i^*} > 0$ and $\sum_{i \in I} \lambda_i \mathbf{d}_i = 0$ for suitable $\lambda_i \geq 0$. Stated otherwise, the homogeneous system of linear equations $\sum_{i \in I} \lambda_i d_{ik} = 0$, $k = 1, \ldots, r$ has nontrivial solutions $\lambda_i \geq 0$. Now, the solutions of this system are obtained by fixing arbitrarily some values, say $\lambda_{\alpha_1}, \ldots, \lambda_{\alpha_p}$, with p strictly smaller than the number of elements of I, and then, for each $\beta \in I \setminus \{\alpha_1, \ldots, \alpha_p\}$, computing λ_β by means of the formulas $\lambda_\beta = \sum_{i=1}^{p} C_{\beta\alpha_i} \lambda_{\alpha_i}$. Here, the coefficients $C_{\beta\alpha_i}$ arise as the result of performing rational operations on the integers d_{ij}. Thus, by continuity, the existence of positive real solutions of the system implies the existence of *rational* positive solutions. Choosing all λ_i rationals and multiplying by their least common denominator, we finally obtain integers $\{n_i \geq 0 \mid i \in I\}$ such that $\sum_{i \in I} n_i \mathbf{d}_i = 0$, and $n_{i^*} \geq 1$. By definition of P, we have $0 \neq \mathbf{d}_{i^*} \leq_P \sum_{i \in I} n_i \mathbf{d}_i = 0$, whence $\mathbf{d}_{i^*} = 0$, a contradiction.

Having thus settled our first claim, it easily follows that

$$(2.25) \qquad P^* \cap N^* = \{0\}$$

From equations (2.20) and (2.25) we obtain $P^* \cap P = P^* \cap \mathbf{Z}^r$, whence in particular, for arbitrary $i, j = 0, \ldots, t$,

$$(2.26) \quad \mathbf{h}_i \leq_P \mathbf{h}_j \ \text{ iff } \ \mathbf{h}_j - \mathbf{h}_i \in P^*.$$

To conclude the proof we need the following well known result of linear algebra, whose proof is included for the sake of completeness. Here, as usual, $\mathbf{h} \cdot \mathbf{k}$ denotes the scalar product of vectors $\mathbf{h}, \mathbf{k} \in \mathbf{R}^r$.

Claim 2: Let $\mathbf{e}_1, \ldots, \mathbf{e}_m$ be vectors in $\mathbf{R}^r$ such that for any sequence of real numbers $0 \leq \lambda_1, \ldots, \lambda_m$, if $\sum_{i=1}^m \lambda_i \mathbf{e}_i = 0$, then $\lambda_1 = \ldots = \lambda_m = 0$. Then there is a vector $\mathbf{v} \in \mathbf{R}^r$ such that $\mathbf{e}_i \cdot \mathbf{v} > 0$ for each $i = 1, \ldots, m$.

The proof is by induction on m. The basis is trivial. For the induction step, let $\mathbf{e}_1, \ldots, \mathbf{e}_{m+1}$ in $\mathbf{R}^r$ have the property that for any real numbers $0 \leq \lambda_1, \ldots, \lambda_{m+1}$, if $\sum_{i=1}^{m+1} \lambda_i \mathbf{e}_i = 0$, then $\lambda_1 = \ldots = \lambda_{m+1} = 0$. A fortiori, for any $0 \leq \mu_1, \ldots, \mu_m$ with $\sum_{i=1}^m \mu_i \mathbf{e}_i = 0$, we must have $\mu_1 = \ldots = \mu_m = 0$. Then by induction hypothesis there is a vector $\mathbf{u} \in \mathbf{R}^r$ such that $\mathbf{e}_i \cdot \mathbf{u} > 0$ for each $i = 1, \ldots, m$. We now argue by cases:

Case 1: $\mathbf{e}_{m+1} \cdot \mathbf{u} > 0$. Then the desired conclusion follows upon letting $\mathbf{v} = \mathbf{u}$.

Case 2: $\mathbf{e}_{m+1} \cdot \mathbf{u} = 0$. Then one gets the desired result by letting $\mathbf{v} = \alpha\mathbf{u} + \mathbf{e}_{m+1}$, where $\alpha = 1 + \max\{\frac{\mathbf{e}_i \cdot \mathbf{e}_{m+1}}{\mathbf{e}_i \cdot \mathbf{u}} \mid i = 1, \ldots, m\}$.

Case 3: $\mathbf{e}_{m+1} \cdot \mathbf{u} < 0$. Then for each $i = 1, \ldots, m$ let the vector $\mathbf{g}_i$ be defined by

$$\mathbf{g}_i = \mathbf{e}_i - \frac{\mathbf{e}_i \cdot \mathbf{u}}{\mathbf{e}_{m+1} \cdot \mathbf{u}} \mathbf{e}_{m+1}.$$

By our assumptions about the vectors $\mathbf{e}_1, \ldots, \mathbf{e}_{m+1}$, whenever $\nu_i \geq 0$ and $\sum_{i=1}^m \nu_i \mathbf{g}_i = 0$ then necessarily, $\nu_1 = \ldots = \nu_m = 0$. By induction hypothesis, there is a vector $\mathbf{t} \in \mathbf{R}^r$ such that $\mathbf{g}_i \cdot \mathbf{t} > 0$ for $i = 1, \ldots, m$. Let $\mathbf{w}$ be defined by

$$\mathbf{w} = \mathbf{t} - \frac{\mathbf{e}_{m+1} \cdot \mathbf{t}}{\mathbf{e}_{m+1} \cdot \mathbf{u}} \mathbf{u}.$$

Then, by direct inspection, $\mathbf{w}$ satisfies the inequalities $\mathbf{e}_{m+1} \cdot \mathbf{w} = 0$ and $\mathbf{e}_i \cdot \mathbf{w} > 0$ for each $i = 1, \ldots, m$. Proceeding now as in Case 2 we get the desired conclusion, and the claim is settled.

It follows from Claims 1 and 2 that there is a vector

$$\mathbf{g} = (\gamma_1, \ldots, \gamma_r) \in \mathbf{R}^r$$

such that $\mathbf{g} \cdot \mathbf{d}_j > 0$ for all nonzero vectors $\mathbf{d}_j$, $j = 1, \ldots, t$. By continuity, $\mathbf{g}$ can be assumed to be *in general position*, in the sense that $\gamma_1, \ldots, \gamma_r$ are linearly independent over $\mathbf{Q}$. Let

$$\pi^+_{\mathbf{g}} = \{\mathbf{h} \in \mathbf{R}^r \mid \mathbf{h} \cdot \mathbf{g} \geq 0\}, \quad P' = \pi^+_{\mathbf{g}} \cap \mathbf{Z}^r.$$

Then from (2.24) it follows that

(2.27) $P^* \subseteq \pi^+_{\mathbf{g}}$ and $N^* \subseteq -\pi^+_{\mathbf{g}}$.

Let us now focus attention on the totally ordered abelian group $T' = (\mathbf{Z}^r, \leq_{P'})$. Although T' need not coincide with $T = (\mathbf{Z}^r, \leq_P)$, still by (2.25)-(2.27), for all $i, j = 0, \ldots, t$ such that $\mathbf{h}_i \neq \mathbf{h}_j$ we have

(2.28) $\mathbf{h}_i \leq_P \mathbf{h}_j$ iff $\mathbf{h}_j - \mathbf{h}_i \in P^*$ iff $\mathbf{h}_j - \mathbf{h}_i \notin N^*$ iff $\mathbf{h}_i \leq_{P'} \mathbf{h}_j$.

For any vectors $\mathbf{k}_0, \ldots, \mathbf{k}_n \in \mathbf{Z}^r$, the map $y \mapsto \mathbf{k}_0, x_1 \mapsto \mathbf{k}_1, \ldots, x_n \mapsto \mathbf{k}_n$ uniquely extends to an interpretation $\sigma_j \mapsto \mathbf{k}_j$, $j = 0, \ldots, t$ of all subterms σ_j of $\hat{\tau}$ into elements $\mathbf{k}_j$ of T'. In the particular case when $\mathbf{k}_0 = \mathbf{h}_0, \ldots, \mathbf{k}_n = \mathbf{h}_n$, arguing by induction on the number of operation symbols occurring in σ_j, from (2.28) we obtain $\mathbf{k}_j = \mathbf{h}_j$ for all $j = 0, \ldots, t$; moreover, all inequalities in (2.21) are still valid with respect to the new total order relation $\leq_{P'}$ over $\mathbf{Z}^r$. In symbols, $0 \leq_{P'} \mathbf{h}_1, \ldots, \mathbf{h}_n \leq_{P'} \mathbf{h}_0$, and $0 \leq_{P'} \mathbf{h}_t \leq_{P'} \mathbf{h}_0$, $0 \neq \mathbf{h}_t = \hat{\tau}^{T'}(\mathbf{h}_1, \ldots, \mathbf{h}_n, \mathbf{h}_0)$.

As an effect of the independence of the γ's over $\mathbf{Q}$, T' is isomorphic, as a totally ordered group, to the subgroup $U = \mathbf{Z}\gamma_1 + \ldots + \mathbf{Z}\gamma_r$ of $\mathbf{R}$ generated by $\gamma_1, \ldots, \gamma_r$, with the natural order. An isomorphism is given by the map

$$\theta\colon \mathbf{b} = (b_1, \ldots, b_r) \in T' \mapsto \mathbf{b} \cdot \mathbf{g} = b_1\gamma_1 + \cdots + b_r\gamma_r \in U.$$

Since the inequalities in (2.21) are preserved under isomorphism, letting $\kappa_0 = \theta(\mathbf{h}_0), \kappa_1 = \theta(\mathbf{h}_1), \ldots, \kappa_n = \theta(\mathbf{h}_n), \kappa_t = \theta(\mathbf{h}_t)$ we can write $0 \leq \kappa_1, \ldots, \kappa_n \leq \kappa_0$ and $0 < \kappa_t \leq \kappa_0$. Assuming without loss of generality, $\kappa_0 = 1$, we have $\kappa_t = \hat{\tau}^U(\kappa_1, \ldots, \kappa_n, 1) > 0$. By Proposition 2.5.1, in the MV-algebra $B = \Gamma(U, 1)$ we have $\tau^B(\kappa_1, \ldots, \kappa_n) \neq 0$, whence, a fortiori, the equation $\tau = 0$ fails in the MV-algebra $[0, 1]$. $\square$

2.6 Bibliographical remarks

The idea of associating a totally ordered abelian group to any MV-algebra A is due to Chang, who in [36] and [38] gave the first purely algebraic proof of the completeness of the Łukasiewicz axioms for the infinite-valued calculus, using quantifier elimination for totally ordered divisible abelian groups.

After the unpublished proof of Wajsberg, the literature contains many other proofs of the completeness theorem: a proof of Rose and Rosser, based on syntactic methods and linear inequalities [216], the proof in [50] using the representation of free ℓ-groups, the proof of Panti [197] using techniques from algebraic geometry. For the one-variable fragment the reader may also see [193]. Our present geometric proof, using elementary algebra and convexity theory in finite-dimensional vector spaces, was first published in [51] (also see [52] for its counterpart for lattice-ordered abelian groups). Claim 2 in the proof is a variant of Farkas' lemma (see [250]). Good sequences and the $\mathbf{\Gamma}$ functor were first introduced in [163].

Chapter 3

Free MV-algebras

Free algebras are universal objects: every n-generated MV-algebra A is a homomorphic image of the free MV-algebra $Free_n$ over n generators; if an equation is satisfied by $Free_n$ then the equation is automatically satisfied by all MV-algebras. As a consequence of the completeness theorem, $Free_n$ is easily described as an MV-algebra of piecewise linear continuous $[0, 1]$-valued functions defined over the cube $[0, 1]^n$. Known as McNaughton functions, they stand to MV-algebras as $\{0, 1\}$-valued functions stand to boolean algebras. Many interesting classes of MV-algebras can be described as algebras of $[0, 1]$-valued continuous functions over some compact Hausdorff space. The various representation theorems presented in this chapter all depend on our concrete representation of free MV-algebras.

3.1 McNaughton functions

Let κ be an arbitrary, finite or infinite cardinal ≥ 1. (Readers not familiar with ordinals and cardinals may think of κ as being a positive integer, without any essential loss.)

Suppose we are given distinct propositional variables

$$X_0, X_1, \ldots, X_\alpha, \ldots$$

for each ordinal $\alpha < \kappa$. Then by definition, each MV-term τ in the variables $\{X_\alpha\}_{\alpha<\kappa}$ is a finite string of symbols over the alphabet

$$\{0, \neg, \oplus, (,), X_\alpha\}_{\alpha<\kappa}$$

obtained by the same inductive procedure of Definition 1.4.1.

For any MV-algebra A the elements of A^κ have the form

$$\mathbf{x} = (x_0, x_1, \ldots, x_\alpha, \ldots)_{\alpha<\kappa},$$

where $x_\alpha \in A$ for each $\alpha < \kappa$. The αth *projection* $\pi_\alpha : A^\kappa \to A$ is the function $\pi_\alpha(\mathbf{x}) = x_\alpha$. We let

$$\mathrm{Proj}_\kappa^A =_{def} \{\pi_\alpha \mid \alpha < \kappa\}$$

denote the set of projections of A^κ.

The following is a generalization of Definition 1.4.3:

Definition 3.1.1 For each term τ in the variables $\{X_\alpha\}_{\alpha<\kappa}$ the *term function*

$\tau^A: A^\kappa \to A$ is given by induction on the number of connectives in τ as follows:

(i) $X_\alpha^A =_{def} \pi_\alpha$n

(ii) $0^A =_{def}$ the constant function 0 over A^κ,

(iii) $(\neg\rho)^A =_{def} \neg(\rho^A)$,

(iv) $(\rho \oplus \sigma)^A =_{def} (\rho^A \oplus \sigma^A)$.

The dependence of τ^A on κ is tacitly understood. We let

$$Term_\kappa^A$$

denote the set of all term functions over A^κ. By construction, each element of $Term_\kappa^A$ is a function only depending on a finite number of variables. Further, $Term_\kappa^A$ is a subalgebra of the MV-algebra A^{A^κ} of all A-valued functions over A^κ, with pointwise defined operations. More precisely, we have the following result, whose proof immediately follows by definition:

Lemma 3.1.2 *For each MV-algebra A and cardinal $\kappa \geq 1$, $Term_\kappa^A$ is the smallest subalgebra of A^{A^κ} containing each projection $\pi_\alpha \in Proj_\kappa^A$.* □

Definition 3.1.3 An MV-algebra A with a distinguished subset Y of elements generating A is said to be *free over (the generating set) Y*, and is denoted by $Free_Y$, iff for every MV-algebra B and every function $f : Y \to B$, f can be extended to a homomorphism $\tilde{f} : A \to B$.

Note that $\tilde{f}$ is uniquely determined by f. A moment's reflection shows that for any two sets Y and Y' of the same cardinality κ, if A is free over Y and A' is free over Y' then $A \cong A'$. Thus there is no danger of ambiguity in saying that A is *the* free MV-algebra over κ many generators, and writing $A = Free_\kappa$.

Proposition 3.1.4 *For each cardinal $\kappa \geq 1$, $Term_\kappa^{[0,1]}$ is the free MV-algebra over the generating set $Proj_\kappa^{[0,1]}$, in symbols, $Term_\kappa^{[0,1]} = Free_\kappa$.*

Proof: Let B be an arbitrary MV-algebra, and $f : Proj_\kappa^{[0,1]} \to B$ be a function, with the intent of uniquely extending f to a homomorphism $\tilde{f} : Term_\kappa^{[0,1]} \to B$. Let $\mathbf{b} = (b_0, b_1, \ldots, b_\alpha, \ldots)_{\alpha<\kappa} \in B^\kappa$ be given by $b_\alpha =_{def} f(\pi_\alpha)$. Let the function φ map each term τ (in the variables X_α) into the element $\tau^B(\mathbf{b}) \in B$, where $\tau^B \in Term_\kappa^B$ is the term function on B^κ determined by τ. By Definition 1.4.4 together with the Completeness Theorem 2.5.3, for any two terms ρ and σ we have

$$\text{if } [0,1] \models \rho = \sigma \text{ then } B \models \rho = \sigma \text{ and then } \rho^B(\mathbf{b}) = \sigma^B(\mathbf{b}).$$

Stated otherwise, whenever $\rho^{[0,1]} = \sigma^{[0,1]} \in Term_\kappa^{[0,1]}$, then φ maps ρ and σ into the same element of B. Since $\varphi(X_\alpha) = X_\alpha^B(\mathbf{b}) = b_\alpha = f(\pi_\alpha)$, then φ naturally determines an extension $\tilde{f}$ of f. By induction one easily sees that $\tilde{f}$ is indeed a homomorphism of $Term_\kappa^{[0,1]}$ into B. To prove uniqueness, let $g : Term_\kappa^{[0,1]} \to B$ be a homomorphism extending f. Since $\tilde{f}$ and g coincide over a subalgebra of A^{A^κ} containing all projections, by Lemma 3.1.2 they coincide over all of $Term_\kappa^{[0,1]}$. □

The proof of the following proposition is now an immediate consequence of Lemma 1.2.3 and Theorem 1.2.8:

Proposition 3.1.5 *Let $\kappa \geq 1$ be a cardinal. Suppose the MV-algebra A is generated by $\leq \kappa$ elements. Then there is an ideal J in the free MV-algebra $Free_\kappa$ such that A is isomorphic to the quotient algebra $Free_\kappa/J$.* $\Box$

Our next task is to give a more explicit description of the elements of the algebras $Free_\kappa$.

Definition 3.1.6 Let $n \geq 1$ be an integer. Then a function

$$f : [0,1]^n \to [0,1]$$

is called a *McNaughton function* over $[0,1]^n$ iff it satisfies the following conditions:

(i) f is continuous with respect to the natural topology of $[0,1]^n$;

(ii) there are linear polynomials $p_1, \ldots, p_k$ with integer coefficients,

$$p_i(x_0, \ldots, x_{n-1}) = b_i + m_{i0}x_0 + \ldots + m_{i(n-1)}x_{n-1},$$

$(b_i, m_{it} \in \mathbf{Z})$, such that for each point $\mathbf{y} = (y_0, \ldots, y_{n-1}) \in [0,1]^n$ there is an index $j \in \{1, \ldots, k\}$ with $f(\mathbf{y}) = p_j(\mathbf{y})$.

Definition 3.1.7 Let λ be an infinite cardinal. A function

$$g : [0,1]^\lambda \to [0,1]$$

is a *McNaughton function* over $[0,1]^\lambda$ iff there are ordinals $\alpha(0) < \ldots < \alpha(m-1) < \lambda$ and a McNaughton function f over $[0,1]^m$ such that for each $\mathbf{x} \in [0,1]^\lambda$, $g(\mathbf{x}) = f(x_{\alpha(0)}, \ldots, x_{\alpha(m-1)})$.

Proposition 3.1.8 *If a function f belongs to a free MV-algebra $Free_\kappa$, then f is a McNaughton function.*

Proof: Trivially, the projections are McNaughton functions, and so is the function constantly taking the value 0 over $[0,1]^\kappa$. The set of McNaughton functions is closed under pointwise application of the $\oplus$ and $\neg$ operations. To see that McNaughton functions are closed under the operation $\oplus$, if f and g are given by linear polynomials $p_1, \ldots, p_m$ and $q_1, \ldots, q_n$, then $f \oplus g$ is given by the linear polynomials $p_i + q_j$, (for all $i = 1, \ldots, m$ and $j = 1, \ldots, n$,) together with the constant function 1. We conclude that the McNaughton functions form a subalgebra of $[0,1]^{[0,1]^{\kappa}}$. By Lemma 3.1.2, all term functions are McNaughton functions. Now apply Proposition 3.1.4. □

McNaughton's theorem, to be proved in a subsequent chapter, states the converse of Proposition 3.1.8. A short proof of the one-variable case shall be given in the next section. For most applications, however, one does not need the full strength of McNaughton's theorem, but only Lemma 3.1.9 below—a much simpler result.

For each real-valued function γ, we let

(3.1) $\gamma^{\sharp} =_{def} (\gamma \vee 0) \wedge 1.$

Lemma 3.1.9 *Let $g : [0,1]^n \to \mathbf{R}$ be a linear function with integer coefficients, say, $g(\mathbf{x}) = m_0x_0 + \ldots + m_{n-1}x_{n-1} + b$, with $m_0, \ldots, m_{n-1}, b \in \mathbf{Z}$. Then $g^{\sharp} \in Free_n$.*

Proof: By induction on $m = |m_0| + \ldots + |m_{n-1}|$. If $m = 0$, then $g^{\sharp}$ coincides with either constant function 0 or 1, whence it trivially belongs to $Free_n$. For the induction step, assume the lemma holds for $m - 1$. It is no loss of generality to assume that $\max(|m_0|, \ldots, |m_{n-1}|) = |m_0|$. Proceeding by cases, assume first $m_0 > 0$. Let $h = g - x_0$, so that

(3.2) $h = h(x_0, \ldots, x_{n-1}) = b + (m_0 - 1)x_0 + m_1x_1 + \ldots + m_{n-1}x_{n-1}.$

By induction hypothesis, both functions $h^{\sharp}$ and $(h+1)^{\sharp}$ belong to $Free_n$. We shall prove that for each $\mathbf{x} = (x_0, \ldots, x_{n-1}) \in [0,1]^n$,

(3.3) $(h + x_0)^{\sharp} = (h^{\sharp} \oplus x_0) \odot (h+1)^{\sharp}.$

The identity trivially holds whenever $\mathbf{x}$ is such that $h(\mathbf{x}) > 1$, or $h(\mathbf{x}) < -1$. If $\mathbf{x}$ is such that $h(\mathbf{x}) \in [0,1]$, then $h^{\sharp}(\mathbf{x}) = h(\mathbf{x})$, and $(h(\mathbf{x})+1)^{\sharp} = 1$. Since $x_0 \in [0,1]$, $(h(\mathbf{x}) + x_0)^{\sharp} = h(\mathbf{x}) \oplus x_0$, which establishes (3.3). Finally, if $h(\mathbf{x}) \in [-1,0]$, then $h^{\sharp}(\mathbf{x}) = 0$ and $(h(\mathbf{x}) + 1)^{\sharp} = h(\mathbf{x}) + 1$, whence (3.3) follows from the identities $(h(\mathbf{x})+x_0)^{\sharp} = \max(0, h(\mathbf{x})+x_0) = \max(0, x_0 + h(\mathbf{x}) + 1 - 1) = x_0 \odot (h(\mathbf{x}) + 1)$. Thus, in case $m_0 > 0$, identity (3.3) holds for each $\mathbf{x} = (x_0, \ldots, x_{n-1})$. By induction, together with Proposition 3.1.4, we have $(h+x_0)^\sharp = g^\sharp \in Free_n$. In case $m_0 < 0$, applying the same argument to the function $1 - g$, one shows that $(1-g)^\sharp \in Free_n$. Since $1 - (1-g)^\sharp = g^\sharp$, we conclude that $g^\sharp \in Free_n$ in all possible cases. □

3.2 The one-dimensional case

In this section we shall prove that, up to isomorphism, the free MV-algebra $Free_1$ over one generator coincides with the MV-algebra of one-variable McNaughton functions.

To this purpose, for each $n = 0, 1, \ldots,$ the nth *Farey partition* Farey_n of the unit interval $[0,1]$ is defined by

$$\begin{array}{lll} \mathrm{Farey}_0 & =_{def} & \{0, 1\}, \\ \mathrm{Farey}_1 & =_{def} & \{0, \frac{1}{2}, 1\}, \\ \mathrm{Farey}_2 & =_{def} & \{0, \frac{1}{3}, \frac{1}{2}, \frac{2}{3}, 1\}, \\ \mathrm{Farey}_3 & =_{def} & \{0, \frac{1}{4}, \frac{1}{3}, \frac{2}{5}, \frac{1}{2}, \frac{3}{5}, \frac{2}{3}, \frac{3}{4}, 1\}, \\ \mathrm{Farey}_4 & =_{def} & \{0, \frac{1}{5}, \frac{1}{4}, \frac{2}{7}, \frac{1}{3}, \frac{3}{8}, \frac{2}{5}, \frac{3}{7}, \frac{1}{2}, \frac{4}{7}, \frac{3}{5}, \frac{5}{8}, \frac{2}{3}, \frac{5}{7}, \frac{3}{4}, \frac{4}{5}, 1\}, \\ \ldots & \ldots & \ldots \end{array}$$

Thus, Farey_{n+1} is obtained by inserting between any two consecutive elements a/b and c/d of Farey_n their *mediant* $(a+c)/(b+d)$, with $0 = 0/1$ and $1 = 1/1$. Our present Farey_n is a notationally simpler variant of the traditional "Farey sequence" of order $n+1$, where one inserts only those mediants whose denominators do not exceed the value $n+1$. Farey observed, and Cauchy proved, the following elementary properties of Farey partitions:

Proposition 3.2.1 *Let $n = 0, 1, 2, \ldots$. Then we have*

(i) All fractions in Farey_n are automatically in irreducible form, and the interval $[a/b, c/d]$ determined by any two consecutive fractions $a/b < c/d$ in Farey_n has the unimodularity *property $cb - ad = 1$. Moreover, $a/b < (a+c)/(b+d) < c/d$.*

(ii) Every irreducible fraction $p/q \in [0,1]$ occurs in Farey$_m$, for some index m.

Proof: (i) Let $(1,1)$ and $(0,1)$ be the vectors in $\mathbf{Z}^2$ respectively giving the *homogeneous correspondents* of 1 and 0. Then the fact that the closed interval $[0/1, 1/1]$ satisfies unimodularity has the following equivalent reformulations:
(a) the determinant of the matrix whose rows are given by $(1,1)$ and $(0,1)$ is equal to one;
(b) every vector in $\mathbf{Z}^2$ is a linear combination, with integer coefficients, of the vectors $(1,1)$ and $(0,1)$—for short, the pair $((1,1),(0,1))$ is a *basis* in $\mathbf{Z}^2$.

One then immediately sees that if an interval $[u/v, u'/v'] \subseteq [0,1]$ satisfies the unimodular law, then so do the two intervals $[u/v, u''/v'']$ and $[u''/v'', u'/v']$ where $u''/v'' =_{def} (u+u')/(v+v')$. For short, unimodularity is preserved under the operation of taking mediants. The rest is clear.

To prove (ii), by way of contradiction assume $0 < p/q < 1$ to be an irreducible fraction not occurring in any Farey$_n$. The coordinates of vector $(p,q) \in \mathbf{Z}^2$ in the basis $B_0 =_{def} ((1,1),(0,1))$ are given by p and $q-p$. Note that their sum is strictly less than the sum $p+q$ of the coordinates of the same vector in the initial basis $((1,0),(0,1))$. Taking the mediant $1/2$ of $0/1$ and $1/1$ (the latter two fractions giving the *inhomogeneous correspondents* of 0 and 1), and passing to homogeneous coordinates in $\mathbf{Z}^2$, we obtain two new bases $((1,1),(1,2))$ and $((1,2),(0,1))$. Precisely one of them, denoted B_1, *encapsulates* the vector (p,q), in the sense that the coordinates α_1 and β_1 of (p,q) with respect to basis B_1 are integers ≥ 1. By direct inspection, the sum of these coordinates is strictly less than q. Proceeding inductively, assume we are given an encapsulating basis $B_j = (\mathbf{v}_j, \mathbf{w}_j)$, for suitable integer vectors $\mathbf{v}_j$ and $\mathbf{w}_j$ in $\mathbf{Z}^2$, and let α_j and β_j be the coordinates of (p,q) with respect to B_j. Upon taking the mediant of the inhomogeneous correspondents of $\mathbf{v}_j$ and $\mathbf{w}_j$, we obtain two bases $(\mathbf{v}_j, \mathbf{v}_j + \mathbf{w}_j)$ and $(\mathbf{v}_j + \mathbf{w}_j, \mathbf{w}_j)$; precisely one of them, denoted B_{j+1}, encapsulates (p,q); further, the coordinates α_{j+1} and β_{j+1} of (p,q) with respect to B_{j+1} satisfy the inequality $2 \leq \alpha_{j+1} + \beta_{j+1} < \alpha_j + \beta_j$. Thus, by our absurdum hypothesis, the sum of the coordinates of (p,q) decreases infinitely often, which is impossible. □

The $2^n + 1$ elements of Farey_n can be displayed in increasing order as follows:

$$0 < \alpha < \ldots < \gamma < \delta = c/d < \varepsilon < \ldots < \omega < 1.$$

A *(Schauder) hat* of Farey_n is a function of either form:

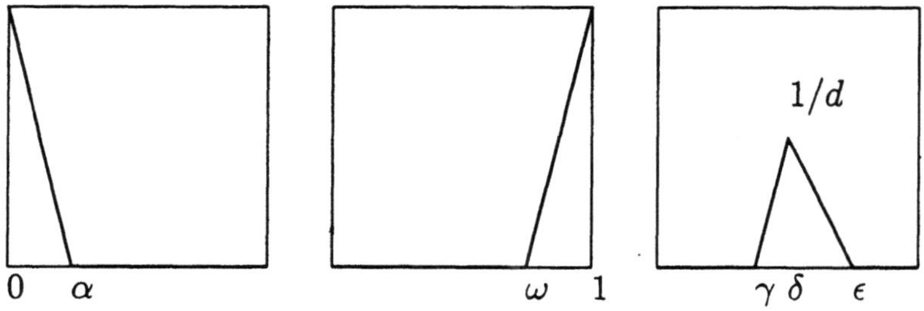

Thus each Schauder hat is a continuous piecewise linear function $h : [0,1] \to [0,1]$, whose graph consists of the four segments joining the points $(0,0)$, $(\gamma,0)$, $(\delta,1/d)$, $(\varepsilon,0)$ and $(1,0)$. As a consequence of unimodularity, each linear piece of h has the form $y = mx + q$, for suitable integers m and q. Of course, the graphs of the two extremal hats only consist of two segments. For each $n = 0,1,2,\ldots,$ we shall denote by

$$\mathrm{Schauder}_n =_{def} (h_1, \ldots, h_u)$$

the naturally ordered sequence of all Schauder hats of Farey_n, where $u = 2^n + 1$. We naturally say that h_j and h_{j+1} are *contiguous*. The following is a picture of $\mathrm{Schauder}_i$, for $i = 0,1,2$:

$\mathrm{Schauder}_0$ $\mathrm{Schauder}_1$ $\mathrm{Schauder}_2$

For each hat h_i of Schauder$_n$ the *multiplicity* μ_i of h_i is the inverse of the maximum value of h_i. Thus by definition, μ_i coincides with the denominator d_i of the rational point $c_i/d_i \in [0,1]$ at which h_i attains its maximum value. For any n, the Schauder hat h_i precedes h_j in Schauder$_n$ iff $i < j$ iff $c_i/d_i < c_j/d_j$. Note that if $|i-j| \geq 2$ then $h_i \wedge h_j = 0$. From the sequence $(h_1, \ldots, h_u)$ one can unambiguously recover the multiplicity of each hat, as well as the number $n = \log_2(u-1)$. It is easy to see that $\sum_i \mu_i h_i = 1$.

By a *node* of a continuous function $f : [0,1] \to [0,1]$ we mean a nondifferentiability point in the domain of f. The vertices 0 and 1 are always included among the nodes of f. If f arises from the identity function via a finite number of applications of the operations $\neg$, $\oplus$, $\odot$, $\wedge$ and $\vee$, then the nodes of f are finitely many, and they are all rational numbers. By Proposition 3.1.4 this is always the case when $f \in Free_1$.

Proposition 3.2.2 *Let $h_1, \ldots, h_u$ be the hats of Schauder$_n$, in their natural order, and with their respective multiplicities $\mu_1, \ldots, \mu_u$. Let $k_1, \ldots, k_{2u-1}$ be the hats of Schauder$_{n+1}$, with their respective multiplicities $\xi_1, \ldots, \xi_{2u-1}$. Then we have:*

$$k_1 = h_1 - (h_1 \wedge h_2) = h_1 \ominus (h_1 \wedge h_2), \text{ with } \xi_1 = \mu_1 = 1;$$

$$k_{2u-1} = h_u - (h_{u-1} \wedge h_u) = h_u \ominus (h_{u-1} \wedge h_u), \text{ with } \xi_{2u-1} = \mu_u = 1;$$

$$k_{2i} = h_i \wedge h_{i+1}, \text{ with } \xi_{2i} = (\mu_i + \mu_{i+1}), \text{ for each } i = 1, 2, \ldots, u-1;$$

$$k_{2i-1} = h_i - (h_i \wedge (h_{i-1} \vee h_{i+1})) = h_i - (h_i \wedge (h_{i-1} + h_{i+1})) = h_i \ominus (h_{i-1} \oplus h_{i+1}), \text{ with } \xi_{2i-1} = \mu_i, \text{ for each } i = 2, 3, \ldots, u-1.$$

Proof: Let $0 = c_1/d_1 < c_2/d_2 < \ldots < c_u/d_u = 1$ be the ascending sequence of elements of Farey$_n$. Then $d_i = \mu_i$. Direct inspection, in the light of the unimodularity property in Proposition 3.2.1 (i), shows that the function $h_i \wedge h_{i+1}$ attains its maximum value $1/(d_i + d_{i+1})$ at the mediant point $\rho = (c_i + c_{i+1})/(d_i + d_{i+1})$. Since the nodes of $h_i \wedge h_{i+1}$ are given by the five rational numbers 0, c_i/d_i, ρ, c_{i+1}/d_{i+1}, 1, and the function coincides with k_{2i} at each node, then k_{2i} coincides with $h_i \wedge h_{i+1}$ over all of $[0,1]$. Similarly, the function $h_i - (h_i \wedge (h_{i-1} + h_{i+1}))$ is constantly equal to zero over both intervals $[0, (c_{i-1} + c_i)/(d_{i-1} +$

$d_i)]$ and $[(c_{i+1} + c_i)/(d_{i+1} + d_i), 1]$, and is linear over both intervals $[(c_{i-1} + c_i)/(d_{i-1} + d_i), c_i/d_i]$ and $[c_i/d_i, (c_{i+1} + c_i)/(d_{i+1} + d_i)]$. Since $(h_{i-1} \wedge h_{i+1}) = 0$, the function coincides with k_{2i-1} at the point c_i/d_i, where it attains its maximum value $1/d_i$. It is now easy to see that the identity $k_{2i-1} = h_i - (h_i \wedge (h_{i-1} + h_{i+1}))$ holds over all of $[0, 1]$. One similarly proves the remaining identities. □

By direct inspection we have

Proposition 3.2.3 *(i) Each one of the extremal hats h_1 and h_u of Schauder$_n$ is the sum of two hats of Schauder$_{n+1}$; all the remaining hats of Schauder$_n$ are obtainable as a sum of three hats of Schauder$_{n+1}$.*
(ii) Suppose a linear combination $\lambda_1 h_1 + \ldots + \lambda_u h_u$ with real coefficients λ_i coincides with the constant function 0. Then each coefficient λ_i must vanish. For short, the hats of Schauder$_n$ form a linearly independent set of functions. □

Definition 3.2.4 Fix an integer $n = 0, 1, \ldots$, and let $h_1, \ldots, h_u$ be the hats of Schauder$_n$, with their respective multiplicities $\mu_1, \ldots, \mu_u$. Then by a *subsystem* S of Schauder$_n$ (in symbols, and with a slight abuse of notation, $S \subseteq$ Schauder$_n$) we mean a sequence of integers $\lambda_1, \ldots, \lambda_u$, with $0 \leq \lambda_i \leq \mu_i$ for each $i = 1, \ldots, u$. The *associated function* $f_S : [0, 1] \to [0, 1]$ is defined by $f_S = \sum_i \lambda_i h_i$. We further let $S^\neg \subseteq$ Schauder$_n =_{def} (\mu_1 - \lambda_1, \ldots, \mu_u - \lambda_u)$, whence, recalling that $\mu_1 h_1 + \ldots + \mu_u h_u = 1$, we have the identity $f_{S^\neg} = \neg f_S = 1 - f_S$. Two subsystems $S \subseteq$ Schauder$_n$ and $R \subseteq$ Schauder$_m$ are said to be *equivalent* iff $f_R = f_S$.

From Proposition 3.2.3 we get

Proposition 3.2.5 *If $S \subseteq$ Schauder$_n$ then for each integer $p \geq n$ there is precisely one subsystem $S' \subseteq$ Schauder$_p$ equivalent to S.* □

Proposition 3.2.6 *Given two subsystems $R \subseteq$ Schauder$_m$ and $S \subseteq$ Schauder$_n$, there exists an integer $q \geq \max(m, n)$ and a subsystem $T \subseteq$ Schauder$_q$ such that $f_T = f_R \oplus f_S$.*

Proof: Let q be the smallest integer $\geq \max(m, n)$ such that all the nodes of the function $f_R \oplus f_S = \min(1, f_R + f_S)$ occur in Farey$_q$. The existence of q follows from Proposition 3.2.1 (ii), together with Proposition 3.2.2 ensuring that all the nodes of $f_R \oplus f_S$ are rational numbers. Let $g = f_R + f_S$. For any two consecutive fractions x and y in Farey_q one cannot have $g(x) > 1$ and $g(y) < 1$. Let $R' = (\lambda_1, \dots, \lambda_w)$ and $S' = (\nu_1, \dots, \nu_w)$ be the equivalent subsystems of S and of R in Schauder_q, as given by Proposition 3.2.5. Let $k_1, \dots, k_w$ be the hats in Schauder_q, with their multiplicities $\xi_1, \dots, \xi_w$. Thus, for any two consecutive indexes $i, j \in \{1, \dots, w\}$, if $\lambda_j + \nu_j < \xi_j$ then necessarily $\lambda_i + \nu_i \le \xi_i$. Let $T \subseteq \text{Schauder}_q$ be the subsystem $(\eta_1, \dots, \eta_w)$, where $\eta_i = \min(\xi_i, \lambda_i + \nu_i)$. It follows that the function f_T coincides with $f_R \oplus f_S$ at each point of Farey_q, and both functions are linear on each interval between consecutive points of Farey_q. Thus, $f_T = f_R \oplus f_S$ as required. □

Example: With the notation of Proposition 3.2.6, let $m = n = 0$ and $R = S = (0, 1)$, so that $\text{Schauder}_m = (1 - x, x)$ and $f_R = f_S = x$. The nodes of the function $f_R \oplus f_S = \min(2x, 1)$ are 0, 1/2 and 1, which all occur in the Farey partition Farey_q for $q = 1$. By definition,

$$\text{Schauder}_1 = (\max(1 - 2x, 0),\ \min(x, 1 - x),\ \max(0, 2x - 1)),$$

with multiplicities 1, 2, 1, respectively. As a particular case of Proposition 3.2.5, the subsystem $R' \subseteq \text{Schauder}_1$ given by the triplet $(0, 1, 1)$ is equivalent to R. Indeed, $f_R = f_{R'} = x$. Letting the subsystem T of Schauder_1 be defined by $T = (0, 2, 1)$, we have $f_R \oplus f_S = f_T$. For no subsystem U of Schauder_0 we would be able to write $f_R \oplus f_S = f_U$.

Theorem 3.2.7 *Let $f : [0, 1] \to [0, 1]$ be an arbitrary function. Then the following are equivalent:*
(i) f arises from the identity function $x : [0, 1] \to [0, 1]$ via a finite number of applications of the operations $\neg$ and $\oplus$;
(ii) there is a subsystem T such that $f = f_T$.

Proof: One direction immediately follows from Proposition 3.2.2, since both operations $\odot$ and $\wedge$ are definable in terms of $\neg$ and $\oplus$. For the other direction, as shown by the above example, the identity function x arises as f_R for some subsystem R. Proceeding by induction on the number of operation symbols $\neg$ and $\oplus$ needed to obtain f, in case $f = \neg g$ one simply notes that $f_{S^\neg} = \neg(f_S)$. In the remaining case, from Propositions 3.2.5 and 3.2.6 it follows that the set of functions of the form f_T, for T a subsystem T of some Schauder$_n$ is closed under applications of $\neg$ and $\oplus$. This yields the desired conclusion. □

Corollary 3.2.8 *The free MV-algebra Free$_1$ over one generator is isomorphic to the MV-algebra of McNaughton functions of one variable.*

Proof: By Proposition 3.1.4, $Free_1$ is a subalgebra of the MV-algebra of one-variable McNaughton functions. Conversely, given any such function f, let us display its nodes as follows:

$$0 = n_1 < n_2 < \ldots < n_{u-1} < n_u = 1.$$

As already remarked, every n_i is a rational number, and f is linear over each interval $[n_i, n_{i+1}]$. By Proposition 3.2.1(ii), there exists an integer q such that all n_i's are members of Farey$_q$. Let

$$0 = m_1 < m_2 < \ldots < m_{v-1} < m_v = 1$$

display the elements of Farey$_q$. Note that the value of f at each node $m_j = a_j/b_j$ is a multiple of $1/b_j$, for some integer $0 \leq t \leq b_j$. It follows that a suitable linear combination g (with integer coefficients) of the hats in Schauder$_q$ will coincide with f at all nodes of m_j. Since both g and f are linear over each interval $[m_i, m_{i+1}]$, then g and f will coincide over the interval $[0, 1]$. There remains to be shown that every Schauder hat is an element of $Free_1$. This is an immediate consequence of Theorem 3.2.7 and Proposition 3.1.4. □

3.3 Decomposing McNaughton functions

In this section, for later use, we discuss some properties of McNaughton functions of n variables.

Let $f : [0, 1]^n \to [0, 1]$ be a McNaughton function. Recalling Definition 3.1.6, let $p_1, \ldots, p_k$ be the *linear constituents* of f, i.e., the distinct polynomials representing the linear pieces of f. For any two such polynomials p and q, either $p \leq q$, or $q \leq p$, or there is a hyperplane H dividing $\mathbf{R}^n$ into two closed half-spaces H^+ and H^- such that $p(\mathbf{x}) \leq q(\mathbf{x})$

holds for each point $\mathbf{x} \in H^+$, while $q(\mathbf{x}) \leq p(\mathbf{x})$ holds for each point $\mathbf{x} \in H^-$. More generally, for every permutation ρ of the set $\{1, \ldots, k\}$, let $P_\rho \subseteq [0,1]^n$ be defined by

$$(3.4) \quad P_\rho =_{def} \{\mathbf{x} \in [0,1]^n \mid p_{\rho(1)}(\mathbf{x}) \geq p_{\rho(2)}(\mathbf{x}) \geq \ldots \geq p_{\rho(k)}(\mathbf{x})\}.$$

As an intersection of the cube $[0,1]^n$ with a finite set of closed half-spaces, each P_ρ is, by definition, a (possibly empty) convex compact *polyhedron*.

By definition, the *vertices* of P_ρ are those points of P_ρ that cannot be expressed as nontrivial convex combinations of points of P_ρ. By the *fundamental theorem on polyhedra*, P_ρ can be visualized as the convex hull of the finite set of its vertices. Since all linear constituents of f have integer coefficients, each vertex $\mathbf{v}$ of P_ρ is *rational*, in the sense that the coordinates of $\mathbf{v}$ are rational numbers.

Let us agree to denote by $\mathcal{C}$ the set of n-dimensional polyhedra of the form P_ρ, for some permutation ρ, in the above decomposition of the domain of f. Then $\mathcal{C}$ is a finite set of compact convex n-dimensional polyhedra with rational vertices, having the following additional properties:

(i) the union of all polyhedra in $\mathcal{C}$ coincides with the cube $[0,1]^n$,

(ii) any two polyhedra in $\mathcal{C}$ are either disjoint or they intersect in a common face, and

(iii) for each polyhedron $P \in \mathcal{C}$ there is an index $u = u_P \in \{1, \ldots, k\}$ such that upon restriction to P, the two functions f and p_u coincide.

For further applications, it will be convenient to replace $\mathcal{C}$ with a family of n-dimensional *simplexes* satisfying (i)-(iii). To this purpose, one can simply triangulate every polyhedron in $\mathcal{C}$, generalizing the familiar triangulation of convex polygons in the two-dimensional case. In this way one does not even need to add new vertices.

As an alternative construction, for each $d = 0, \ldots, n$ let us denote by $\mathcal{F}^{(d)}$ the set of d-dimensional faces of polyhedra in $\mathcal{C}$. Thus, for instance, $\mathcal{F}^{(0)}$ is the set of singletons given by the vertices of polyhedra in $\mathcal{C}$, $\mathcal{F}^{(1)}$ is the set of edges of polyhedra in $\mathcal{C}, \ldots, \mathcal{F}^{(n)} = \mathcal{C}$.

For every $j = 2, 3, \ldots, n$ and every polyhedron $P \in \mathcal{F}^{(j)}$ let us select, once and for all, a rational point $\mathbf{b}_P$ in the relative interior of P. In other words, $\mathbf{b}_P \in P \cap \mathbf{Q}^n$ and $\mathbf{b}_P$ does not belong to any

$(j-1)$-dimensional face of P. We now inductively define a sequence $\mathcal{S}^{(0)}, \mathcal{S}^{(1)}, \ldots, \mathcal{S}^{(n)}$ where each $\mathcal{S}^{(j)}$ is a set of j-dimensional simplexes, as follows:

$$\mathcal{S}^{(0)} = \mathcal{F}^{(0)}, \quad \mathcal{S}^{(1)} = \mathcal{F}^{(1)},$$

$$\mathcal{S}^{(2)} = \{[\mathbf{b}_P, F] \mid P \in \mathcal{F}^{(2)}, F \in \mathcal{S}^{(1)}, F \subseteq P\},$$

$$\ldots \qquad \ldots \qquad \ldots$$

$$\mathcal{S}^{(i+1)} = \{[\mathbf{b}_P, F] \mid P \in \mathcal{F}^{(i+1)}, F \in \mathcal{S}^{(i)}, F \subseteq P\},$$

where $[\mathbf{b}_P, F]$ is the simplex whose vertices are the vertices of F together with $\mathbf{b}_P$. Letting $\mathcal{S} = \mathcal{S}^{(n)}$, we have proved:

Proposition 3.3.1 *For any McNaughton function $f : [0,1]^n \to [0,1]$ with its linear constituents $p_1, \ldots, p_k$, there is a set $\mathcal{S}$ of compact n-dimensional simplexes with rational vertices such that*

(i) The union of all simplexes in $\mathcal{S}$ coincides with the cube $[0,1]^n$;

(ii) Any two simplexes in $\mathcal{S}$ are either disjoint or they intersect in a common face;

(iii) For each simplex $W \in \mathcal{S}$ there is an index $u \in \{1, \ldots, k\}$ such that, upon restriction to W, the two functions f and p_u coincide. □

3.4 Ideals in free MV-algebras

The aim of this section is to describe the maximal ideals in free MV-algebras and to give conditions ensuring that an ideal of a free MV-algebra is an intersection of maximal ideals.

Our results essentially depend on the fact that free MV-algebras are algebras of $[0,1]$-valued continuous functions on a compact Hausdorff topological space. Therefore, we shall firstly recall the ideal theory of such function algebras.

As the reader will remember, for every MV-algebra A, $\mathcal{I}(A)$ and $\mathcal{M}(A)$ respectively denote the set of ideals and of maximal ideals of A.

For X an arbitrary nonempty set, let A be a subalgebra of the MV-algebra $[0,1]^X$ of all functions $f: X \to [0,1]$, with pointwise defined operations. For each set $S \subseteq X$, let

$$(3.5) \quad J_S =_{def} \{f \in A \mid f(x) = 0 \text{ for all } x \in S\},$$

be the set of functions vanishing over S. Further, for any ideal J in A let

$$(3.6) \quad V_J =_{def} \bigcap\{f^{-1}(0) \mid f \in J\}$$

be the intersection of the *zerosets* $Z(f) =_{def} f^{-1}(0)$ of all functions $f \in J$. Trivially, $J_S \in \mathcal{I}(A)$; further, $J_\emptyset = A$ and $J_X = \{0\} = \langle 0 \rangle$. For each $x \in X$ we shall write J_x instead of $J_{\{x\}}$.

Lemma 3.4.1 *Let X be a nonempty set and let A be a subalgebra of the MV-algebra $[0,1]^X$. Then for each $x \in X$, $J_x \in \mathcal{M}(A)$.*

Proof: Suppose first that $A = [0,1]^X$. J_x is a proper ideal of A, because the constant function 1 is not among its elements. If $f \in A \setminus J_x$, then $f(x) > 0$, and we can find an integer $n \geq 1$ such that $nf(x) \geq 1$. It follows that

$$\neg nf = 1 - \underbrace{(f \oplus \cdots \oplus f)}_{n \text{ times}} \in J_x,$$

whence, by Proposition 1.2.2, $J_x \in \mathcal{M}(A)$. Now to complete the proof it suffices to apply Proposition 1.2.16 (ii). □

For any topological space X we shall denote by

$$Cont(X)$$

the subalgebra of $[0,1]^X$ given by the *continuous* $[0,1]$-valued functions over X. It is understood that $[0,1]$ is equipped with its natural topology.

Proposition 3.4.2 *Let X be a compact Hausdorff space, and A be a subalgebra of the MV-algebra $Cont(X)$. The map $J \mapsto V_J$ is an inclusion reversing map from the set $\mathcal{I}(A)$ of ideals into the family of closed subsets of X. Moreover, $V_J \neq \emptyset$ for each proper ideal J in A.*

Proof: The continuity of each function $f \in A$ ensures that V_J is a closed subset of X; trivially, the map $J \mapsto V_J$ reverses inclusions. Let J be a proper ideal in A. By way of contradiction, suppose $V_J = \emptyset$. Then by the assumed compactness of X there are finitely many functions $f_1, \ldots, f_s \in J$ such that the intersection of their zerosets is empty. Let $f = f_1 \oplus \cdots \oplus f_s$. Then $f \in J$ and the zeroset of f is empty. Since f attains a minimum value > 0, there exists an integer $m \geq 1$ such that $mf(x) > 1$ for all $x \in X$. We conclude that $f \oplus \cdots \oplus f$ (m times), constantly takes the value 1. Thus $1 \in J$, and $J = A$, a contradiction. □

A subalgebra A of $[0,1]^X$ is said to be *separating* iff for any two distinct points x and y in X, there is $f \in A$ such that $f(x) = 0$ and $f(y) > 0$.

Theorem 3.4.3 *Let X be a compact Hausdorff space and A be a separating subalgebra of $Cont(X)$. Then we have*

(*i*) *The map $x \mapsto J_x$ is a one-one correspondence between X and $\mathcal{M}(A)$;*

(*ii*) *For each closed set $S \subseteq X$, $V_{J_S} = S$;*

(*iii*) *For each proper ideal J in A, J_{V_J} is the intersection of all maximal ideals in A containing J.*

Proof: (*i*) By Lemma 3.4.1 together with our assumption about A, the map $x \mapsto J_x$ is a one-one correspondence from X into $\mathcal{M}(A)$. To see that this map is onto $\mathcal{M}(A)$, let J be a maximal ideal of A. By Proposition 3.4.2, V_J is a nonempty closed subset of X. Since for each $y \in V_J$, $J_y \supseteq J$, we conclude that V_J is a singleton, say $V_J = \{x\}$ and $J = J_x$.

(*ii*) Trivially, $S \subseteq V_{J_S}$. To prove the converse inclusion, suppose $z \in X \setminus S$. Since A is separating, for each $y \in S$ we can find $f_y \in A$ such that $f_y(z) = 0$ and $f_y(y) = a_y > 0$. By continuity, there is an open neighborhood U_y of y such that $f_y(x) > b_y = a_y/2$ for each $x \in U_y$. By a standard compactness argument, there are finitely many functions $f_1, \ldots, f_k \in A$, together with real numbers numbers $b_1, \ldots, b_k > 0$ such that, upon defining $f = f_1 \oplus \cdots \oplus f_k$, we can write $f(z) = 0$ and

$f(x) > \min(b_1, \ldots, b_k) > 0$ for each $x \in S$. Hence for some integer $n \geq 1$, we must have $\neg nf \in J_S$ and $\neg nf(z) = 1$, thus showing that $z \notin V_{J_S}$. This yields the desired conclusion $V_{J_S} \subseteq S$.

(iii) For each $S \subseteq X$, $J_S = \bigcap_{x \in S} J_x$. On the other hand, $J \subseteq J_x$ if and only if $x \in V_J$. The desired result now follows from (i). □

Remark: Let X be a compact Hausdorff space and A be a separating subalgebra of $Cont(X)$. It follows from (ii) in the above theorem that each closed subset of X is an intersection of zerosets of functions in A. Therefore, *the complements of the zerosets of functions in A form an open basis for the topology of X.*

Corollary 3.4.4 *Let X be a compact Hausdorff space, let A be a separating subalgebra of $Cont(X)$, and J be an ideal in A. Then J is an intersection of maximal ideals iff $J = J_{V_J}$. Moreover, the map $U \mapsto J_{X \setminus U}$ is an order isomorphism from the set of proper open subsets of X and the set of ideals of A that are intersection of maximal ideals (both sets being ordered by inclusion). The inverse isomorphism is given by $J \mapsto X \setminus V_J$.* □

Example: Let $f: [0,1] \to [0,1]$ be defined by $f(x) = x\sin(1/x)$, for $0 < x \leq 1$ and $f(0) = 0$; let further $f^+(x) = \max(f(x), 0)$ for each $x \in [0,1]$. Then $\langle f^+ \rangle$ is a proper ideal of $Cont([0,1])$ and it is not hard to verify that $J_{V_{\langle f^+ \rangle}}$ strictly contains $\langle f^+ \rangle$. Therefore, $\langle f^+ \rangle$ is a principal proper ideal of $Cont([0,1])$ which is not an intersection of maximal ideals.

Let X be a compact Hausdorff space, and A a subalgebra of the MV-algebra $[0,1]^X$. For every nonempty subset Y of X, let

$$A|_Y$$

denote the subalgebra of $[0,1]^Y$ given by the restrictions to Y of the functions in A. The map $f \mapsto \rho(f) = f|_Y$ defines a surjective homomorphism $\rho: A \to A|_Y$. In case Y happens to coincide with V_J for some proper ideal J in A, it follows that $Ker(\rho) = J_{V_J}$. Then from Lemma 1.2.7 we immediately obtain

Proposition 3.4.5 *Let X be a compact Hausdorff space, and A be a separating subalgebra of $Cont(X)$. For each $J \in \mathcal{I}(A)$, the map $f/J \mapsto f|_{V_J}$ is an isomorphism from A/J onto $A|_{V_J}$ if and only if J is an intersection of maximal ideals of A.* □

In order to apply the above results to free MV-algebras, we prepare

Lemma 3.4.6 *For each cardinal $\kappa \geq 1$, $Free_\kappa$ is a separating subalgebra of $Cont([0,1]^\kappa)$.*

Proof: In the light of Proposition 3.1.8, we may visualize the elements of $Free_\kappa$ as McNaughton functions on the cube $[0,1]^\kappa$, the latter being equipped with product topology. Therefore, $Free_\kappa$ is a subalgebra of $Cont([0,1]^\kappa)$. In order to show that this subalgebra is separating, let $\mathbf{y} = (y_0, y_1, \ldots)$ and $\mathbf{z} = (z_0, z_1, \ldots)$ be two distinct points of $[0,1]^\kappa$. Assume without loss of generality $y_0 < z_0$. Let r be a rational number such that $y_0 < r < z_0$. Let $p = p(x) = ax + b$ be a linear polynomial with integer coefficients such that $a > 0$ and $r = -b/a$. Then by Proposition 3.1.4 and Lemma 3.1.9, the McNaughton function $f(\mathbf{x}) = p^\sharp(x_0)$ belongs to the free MV-algebra $Free_\kappa$, and, moreover, $f(\mathbf{y}) = 0$ and $f(\mathbf{z}) > 0$. □

As an immediate consequence of the above lemma together with Theorem 3.4.3, we record here the following result, that will find application in subsequent sections.

Proposition 3.4.7 *The map $\mathbf{x} \mapsto J_\mathbf{x}$ is a one-one correspondence between points of $[0,1]^\kappa$ and maximal ideals of $Free_\kappa$. The inverse correspondence is given by $J \in \mathcal{M}(Free_\kappa) \mapsto$ the only point of V_J.* □

The following lemma is of independent interest:

Lemma 3.4.8 *Let $f, g \in Free_\kappa$. Then*

(3.7) $g \in \langle f \rangle$ *iff* $Zg \supseteq Zf$.

Proof: For the nontrivial direction, assuming $Zg \supseteq Zf$, since both g and f only depend on finitely many variables, say $x_1, \ldots, x_n$, we can safely restrict attention to $[0,1]^n$. By a simple adaptation of Proposition 3.3.1 there exists a set $\mathcal{S} = \{T_1, \ldots, T_u\}$ of compact convex n-dimensional simplexes, whose union coincides with $[0,1]^n$, any two T_i and T_j being either disjoint or intersecting in a common face, and with the additional property that both functions f and g are linear over each T_i. Let $\mathbf{v}_{i0}, \ldots, \mathbf{v}_{in}$ be the vertices of the simplex T_i. For each $1 \leq i \leq u$ and $0 \leq j \leq n$ there is an integer $m_{ij} \geq 1$ such that $g(\mathbf{v}_{ij}) \leq m_{ij} f(\mathbf{v}_{ij})$. As a matter of fact, if $f(\mathbf{v}_{ij}) = 0$ then by hypothesis $g(\mathbf{v}_{ij}) = 0$, and we can take $m_{ij} = 1$. If, on the other hand, $f(\mathbf{v}_{ij}) > 0$, then the existence of m_{ij} follows from the archimedean property of the real numbers. Let $m_i = \max(m_{i0}, \ldots, m_{in})$. Since each $\mathbf{x} \in T_i$ is a convex combination of the vertices of T_i, and since both g and f are linear over T_i, we get $g(\mathbf{x}) \leq m_i f(\mathbf{x})$ for each $\mathbf{x} \in T_i$. Let $m = \max(m_1, \ldots, m_u)$. From $[0,1]^n = \bigcup_{i=1}^{u} T_i$, it follows that $g(\mathbf{x}) \leq m f(\mathbf{x})$ for each $\mathbf{x} \in [0,1]^n$. Since $g \leq 1$, we finally obtain $g(\mathbf{x}) \leq \min(1, m f(\mathbf{x})) = f(\mathbf{x}) \oplus \cdots \oplus f(\mathbf{x})$ (m times). In conclusion, $g \in \langle f \rangle$. □

In contrast with the example given after Corollary 3.4.4, we have

Theorem 3.4.9 *Each proper principal ideal of $Free_\kappa$ is an intersection of maximal ideals.*

Proof: An immediate consequence of Corollary 3.4.4, Lemma 3.4.6, together with Lemma 3.4.8. □

In the next section we shall give an example of a nonprincipal ideal of $Free_1$ which is an intersection of maximal ideals. The following example shows that free MV-algebras also contain proper ideals which are not intersections of maximal ideals.

Example: With reference to formula (3.1), for each $k = 1, 2, \ldots$, let the McNaughton function $f_k : [0,1] \to [0,1]$ be defined by

$$f_k(x) = (k - (k+1)x)^\sharp.$$

Let J be the ideal of $Free_1$ generated by the functions $f_1, f_2, \ldots$. By direct inspection we easily obtain $V_J = \{1\}$, and $(1-x)^\sharp \in J_1 \setminus J$. Therefore, $J \neq J_{V_J}$.

3.5 Simple MV-algebras

An MV-algebra is called *simple* iff it has exactly two ideals. In other words, an MV-algebra A is simple if and only if A is nontrivial and $\{0\}$ is its only proper ideal.

Theorem 3.5.1 *For every MV-algebra A the following conditions are equivalent:*

(*i*) *A is simple;*

(*ii*) *A is nontrivial and for every nonzero element $x \in A$ there is an integer $n > 0$ such that $1 = x \oplus \cdots \oplus x$ (n times);*

(*iii*) *A is isomorphic to a subalgebra of* $[0,1]$.

Proof: In the light of Proposition 1.2.2, (ii) states that $\{0\}$ is a maximal ideal of A. Therefore, (i) and (ii) are equivalent. It is obvious that (ii) is satisfied by all subalgebras of the MV-algebra $[0,1]$. Finally, to prove (i) $\rightarrow$ (iii), assume A to be simple. If the cardinality of A is κ, then by Proposition 3.1.5, we can identify A with the quotient MV-algebra $Free_\kappa/J$ for some ideal J of $Free_\kappa$. By Proposition 1.2.10, since A is simple, J must be a maximal ideal of $Free_\kappa$. Therefore, by Proposition 3.4.7, $J = J_{\mathbf{x}}$, for a uniquely determined point $\mathbf{x} \in [0,1]^\kappa$. Applying now Proposition 3.4.5, we obtain that A is isomorphic to the MV-algebra $Free_\kappa|_{\{\mathbf{x}\}} = \pi_{\mathbf{x}}(Free_\kappa)$, where $\pi_{\mathbf{x}}: Free_\kappa \rightarrow [0,1]$ is the projection given by $\pi_{\mathbf{x}}(f) = f(\mathbf{x})$. Hence A is isomorphic to a subalgebra of $[0,1]$.□

Corollary 3.5.2 *Every simple MV algebra A has at most the cardinality of the continuum.* □

To obtain further information about simple MV-algebras, recall that for each integer $n = 2, 3, \ldots$, the n element (Łukasiewicz) chain $\mathbf{Ł}_n$ is defined by

(3.8) $$\mathbf{Ł}_n =_{def} \{0, \frac{1}{n-1}, \frac{2}{n-1}, \ldots, \frac{n-2}{n-1}, 1\}.$$

Proposition 3.5.3 *Let A be a subalgebra of $[0,1]$. Let $A^+ = \{x \in A \mid x > 0\}$, and $a = \inf A^+$ be the infimum of A^+. If $a = 0$ then A is a dense subchain of $[0,1]$. If $a > 0$ then $A = \text{Ł}_n$ for some $n \geq 2$.*

Proof: In case $a = 0$, let $0 < z \leq 1$ and $0 < \epsilon < z/2$. By assumption there is $b \in A^+$ such that $0 < b < \epsilon$. Letting n be the smallest integer such that $nb \geq z$, noting that $n > 2$, it follows that $z - \epsilon < (n-1)b < z$. Thus A is dense. In case $a > 0$, if $a = 1$ then $A = \text{Ł}_2 = \{0,1\}$. If $a < 1$, since A is closed under the operation $1 - x$, we have $a \leq 1/2$ and $a \in A^+$. For otherwise, there exist two elements $x, y \in A^+$ such that $a < x < y < 2a$, whence $a > y - x = y \odot \neg x \in A^+$, a contradiction. Having proved that $a \in A^+$, let m be the unique integer such that $(m-1)a < 1 \leq ma$. Note that $m \geq 2$. Let

$$K = \{0, a, 2a, \ldots, (m-1)a, 1\} \subseteq A.$$

We shall prove that $K = A$. For otherwise, let $x \in A \backslash K$ (absurdum hypothesis). If $(m-1)a < x < 1$ then $\neg x$ would be a nonzero element of A strictly smaller than a, which is impossible. If for some $j = 0, 1, \ldots, m-2$ we have $ja < x < (j+1)a$ then, again, $(j+1)a - x = (j+1)a \odot \neg x$ is a nonzero element of A which is strictly smaller than a, another contradiction. Having thus proved that $K = A$, since $0 < a < 2a < \ldots < (m-1)a < 1$, it follows that $a = 1 - (m-1)a$, whence $a = 1/m$ and $A = \text{Ł}_{(m+1)}$, as required. □

Remark: Let $m, n \geq 2$. Then from the above proof it follows that $\text{Ł}_m \subseteq \text{Ł}_n$ iff for some $k \in \{1, \ldots, n-1\}$ we have the identity $\frac{1}{m-1} = \frac{k}{n-1}$ iff $m - 1$ is a divisor of $n - 1$.

Corollary 3.5.4 *An MV-algebra A is finite and simple if and only if A is isomorphic to an MV-algebra Ł_n for some integer $n \geq 2$.*

Proof: For the nontrivial direction, if A is finite and simple, by Theorem 3.5.1 we can identify A with a finite subalgebra B of $[0,1]$. By Proposition 3.5.3, letting $n \geq 2$ be the number of elements of B, we have that $B = \text{Ł}_n$. □

As promised, we shall now show that there are maximal ideals in $Free_1$ that are not principal.

Example: Consider the subalgebra S_λ of $[0,1]$ generated by an irrational number $\lambda \in [0,1]$. Since S_λ is simple and has one generator, we can write $S_\lambda = Free_1/J$ for some maximal ideal J of $Free_1$. Now, J cannot be a principal ideal: for otherwise, (absurdum hypothesis), recalling Proposition 3.4.7 and writing $J = J_x = \langle f \rangle$, for some $x \in [0,1]$, and $f \in Free_1$, we get that the zeroset Zf coincides with the singleton $\{x\}$. On the other hand, since all linear pieces of f have integer coefficients, x must be a rational number, say $x = a/b$ for relatively prime integers $0 \leq a \leq b$ with $b > 0$. It follows that $Free_1/J$ is isomorphic to the MV-algebra $Ł_{b+1}$ of possible values of McNaughton functions g at the rational point a/b. Since by Proposition 3.5.3, S_λ is infinite and $Ł_{b+1}$ is finite, these two algebras are not isomorphic. Thus J is nonprincipal.

3.6 Semisimple MV-algebras

In the light of Corollary 1.2.15, for any MV-algebra A, we call *radical of* A the intersection of all maximal ideals of A. The radical of A will be denoted by

$$Rad(A).$$

An MV-algebra A is said to be *semisimple* iff A is nontrivial and $Rad(A) = \{0\}$. In particular, every simple MV-algebra is semisimple.

In what follows, we will consider only *nontrivial* MV-algebras.

As an immediate consequence of Proposition 1.2.10, for each ideal J of A, the quotient A/J is a simple MV-algebra if and only if J is maximal. Hence, by Theorem 3.5.1, A/J is isomorphic to a subalgebra of $[0,1]$ if and only if J is a maximal ideal of A. The next proposition will be promptly recognized as an immediate consequence of Theorem 1.3.2:

Proposition 3.6.1 *An MV-algebra is semisimple if and only if it is a subdirect product of subalgebras of* $[0,1]$. □

Corollary 3.6.2 *Every free MV-algebra is semisimple.*

Proof: Immediate from Proposition 3.1.4. □

Our next aim is to characterize the elements of the radical.

Definition 3.6.3 An element a in an MV-algebra A is said to be *infinitely small* or *infinitesimal* iff $a \neq 0$ and $na \leq \neg a$ for each integer $n \geq 0$. The set of all infinitesimals in A will be denoted by *Infinit*(A).

Remark: If a is infinitesimal, then the elements na, for $n = 0, 1, 2, \ldots$, form a strictly increasing sequence. Indeed, if for some n, $na = (n+1)a$, then one would have $1 = \neg((n+1)a) \oplus na = \neg a \vee na = \neg a$, i.e. $a = 0$, a contradiction.

Proposition 3.6.4 *For any MV-algebra A, $Rad(A) = Infinit(A) \cup \{0\}$.*

Proof: Suppose $a \notin Rad(A)$. Then there is a maximal ideal of A, say M, such that $a \notin M$. Hence $\langle M \cup \{a\} \rangle$, the ideal generated by a and M, must coincide with A, and by (1.9), there is an integer $n \geq 0$ and an element $z \in M$ such that $1 = na \oplus z$. If $na \leq \neg a$, then one would have $a \leq \neg na \leq z$, whence $a \in M$, a contradiction. Hence a is not infinitesimal, whence $Infinit(A) \subseteq Rad(A)$.

Conversely, assume that $0 < a \in A$ is not infinitesimal. Then there is an integer $m \geq 0$ such that $ma \not\leq \neg a$. Hence $ma \ominus \neg a = ma \odot a > 0$, and by Proposition 1.2.13, there is a prime ideal P of A such that $ma \ominus \neg a \notin P$. By Equation (1.7) we have $\neg a \ominus ma \in P$. By Corollary 1.2.12 there is a maximal ideal M of A such that $P \subseteq M$. Then $\neg(a \oplus ma) = \neg a \ominus ma \in M$. Therefore, $(m+1)a \notin M$, and hence, $a \notin M$. In conclusion, $a \notin Rad(A)$, and $Rad(A) \subseteq Infinit(A) \cup \{0\}$, as required to complete the proof. □

It follows from the above proposition that an MV-algebra A is semisimple if and only if A has no infinitesimals. In particular, by the remark following Definition 3.6.3, each finite MV-algebra A is semisimple; moreover, since A has only a finite number of ideals, A is the subdirect product of a finite family of finite subalgebras of $[0, 1]$. The next proposition gives a sharper result:

Proposition 3.6.5 *An MV-algebra A is finite if and only if A is isomorphic to a finite product of finite chains, in symbols,*

(3.9) $A \cong \mathbf{L}_{d_1} \times \cdots \times \mathbf{L}_{d_u}$, *for some integers* $2 \leq d_1 \leq d_2 \leq \ldots \leq d_u$.

This representation is unique, up to the ordering of factors.

Proof: For the nontrivial direction, assume A is a finite MV-algebra. As noted above, A is semisimple. Let $I_1, \ldots, I_u$ be the list of all distinct maximal ideals of A. Then A is a subdirect product of the MV-algebras $A/I_1, A/I_2, \ldots, A/I_u$. By Corollary 3.5.4, for each $t = 1, \ldots, u$ the quotient algebra A/I_t can be identified with the chain Ł_{d_t}, for some integer $d_t \geq 2$. Let us now define the product MV-algebra P by

$$P = \prod\{\text{Ł}_{d_t} \mid t = 1, \ldots, u\}.$$

For each element $x \in A$ and every ideal I_t the element $x/I_t \in \text{Ł}_{d_t}$ can be identified with a rational number $c/(d_t - 1)$, for some integer $0 \leq c \leq d_t - 1$. Let $\beta : x \in A \mapsto (x/I_1, \ldots, x/I_u) \in P$. Then from the semisimplicity of A it follows that β is injective. To see that β is surjective, it is sufficient to show that every element in P of the form $(0, \ldots, 1/(d_t - 1), \ldots, 0)$ is in the range of β. To this purpose, let a_t be the minimum of all elements $z \in A$ such that $z/I_t = 1/(d_t - 1)$. The existence of a_t is ensured by the assumed finiteness of A. For every maximal ideal I_r of A other than I_t, there is an element $w \in A$ such that $a_t \leq w$ and $w/I_r = 0$. For otherwise, $I_r \subseteq I_t$, thus contradicting the maximality of I_r. We have proved that $\beta(a_t) = (0, \ldots, 1/(d_t - 1), \ldots, 0)$, and hence, A is isomorphic to the product MV-algebra P. To prove uniqueness of the decomposition (3.9), suppose

$$A = \text{Ł}_{b_1} \times \cdots \times \text{Ł}_{b_v}$$

for suitable integers $b_i \geq 2$. Then for each $i = 1, \ldots, v$, the kernel of the projection function $(x_1, \ldots, x_r) \mapsto x_i$ is a maximal ideal I_i of A. Moreover, $A/I_i \cong \text{Ł}_{b_i}$ and $A \cong A/I_1 \times \cdots \times A/I_r$. □

Lemma 3.6.6 *Let A be a (nontrivial) MV-algebra, and J be an ideal of A. Then the quotient algebra A/J is semisimple if and only if J is an intersection of maximal ideals of A.*

Proof: Let $h_J : A \to A/J$ be the natural projection.

Suppose that A/J is a semisimple MV-algebra. If $\{M_i\}_{i \in I}$ denotes the family of all maximal ideals of A/J, then

$$J = h_J^{-1}(\{0\}) = h_J^{-1}(\bigcap_{i \in I} M_i) = \bigcap_{i \in I} h_J^{-1}(M_i).$$

By Proposition 1.2.10, J is an intersection of maximal ideals of A.

Conversely, assume J to be an intersection of maximal ideals of A. Then J is the intersection of *all* the maximal ideals of A containing J. If $\{M_i\}_{i\in I}$ denotes this family, then, by Proposition 1.2.10, $\{h_J(M_i)\}_{i\in I}$ is the family of all maximal ideals of A/J and $h_J(J) = Rad(A/J)$. □

The following result strengthens Proposition 3.6.1:

Theorem 3.6.7 *An MV-algebra A with κ many generators is semisimple if and only if for some nonempty closed subset $X \subseteq [0,1]^\kappa$, A is isomorphic to the MV-algebra of restrictions to X of all functions in $Free_\kappa$.*

Proof: One direction is an immediate consequence of Proposition 3.6.1. The other direction follows at once from Propositions 3.1.5 and 3.4.5 and Lemma 3.6.6. □

Corollary 3.6.8 *For any MV-algebra A the following conditions are equivalent:*

(i) A is semisimple;

(ii) A is isomorphic to a separating MV-algebra of $[0,1]$-valued continuous functions on some nonempty compact Hausdorff space, with pointwise operations. □

The following theorem directly follows from Theorem 3.4.9 and Lemma 3.6.6.

Theorem 3.6.9 *Suppose $A \cong Free_\kappa/J$, with J a principal ideal. Then A is semisimple.* □

3.7 Bibliographical remarks

Our definitions of term and free algebras are particular cases of abstract definitions in Universal Algebra (see, for instance, [25], [104] or [150]).

Lemma 3.1.9 is due to McNaughton [152]. The simplified proof presented here is due to Rose and Rosser [216]. The proof of the one variable McNaughton theorem, based on Farey series and Schauder hats, is taken from [181].

Our Farey_n is a variant, due to Stern and Brocot, of the traditional Farey sequence of order $n+1$, considered in most textbooks in number theory.

For a proof of the fundamental theorem on polyhedra see, for instance, [250] or [82, p.31]. The fact, mentioned at the beginning of Section 3, that every polyhedral complex can be refined to a simplicial complex without adding new vertices is also well known: see, for instance, [82, Theorem III, 2.6]. Proposition 3.3.1 can be found in McNaughton's paper [152].

Lemma 3.4.8 was first proved in [172, Proposition 2.4]. Theorem 3.4.9 is the algebraic counterpart of results of Hay, Wójcicki and Rose that will be presented in the next chapter. The proof given here is taken from [170]. The example after Theorem 3.4.9 is due to Wójcicki [243].

Theorem 3.5.1, Proposition 3.6.1 and Corollary 3.5.4 are due to Chang [36] (see also [13],[144]).

Definition 3.6.3 and Proposition 3.6.4 are due to Rodríguez [212] (see also [132, Proposizione 7]). Proposition 3.6.5 appears in [144], [233], [212].

Chapter 4

Łukasiewicz ∞-valued calculus

Since every MV-term τ is a string of symbols over a finite alphabet, one may naturally consider the following *decision problem*: does there exist an effective procedure (for definiteness, a Turing machine) deciding whether an arbitrary equation $\tau = 1$ holds in all MV-algebras ? More generally, given two terms σ and τ, does there exist an effective procedure to decide whether the McNaughton function determined by σ belongs to the principal ideal determined by τ in the free MV-algebra $Free_\omega$? These are respectively known as the word problem for free MV-algebras, and the word problem for finitely presented MV-algebras.

In this chapter we shall reformulate these problems in purely logical terms, within the infinite-valued sentential *calculus* of Łukasiewicz. We shall regard MV-terms as propositions, valid equations as tautologies, ideals as theories, word problems as decision problems in this calculus—in the traditional sense. Adopting this viewpoint, we shall need an equivalent reformulation of Chang's completeness theorem to the effect that *all* tautologies are obtainable from a certain set of initial tautologies (corresponding to the MV-axioms) by a finite number of applications of modus ponens. Free MV-algebras shall be re-obtained as algebras of propositions up to logical equivalence. Using the results of Chapter 3 we shall finally obtain an effective procedure to decide whether a proposition is a tautology, thus automatically giving a positive solution to the word problem for free MV-algebras. Moreover, the word problem for finitely presented MV-algebras shall be reformulated in terms of logical consequence. We shall give a positive solution to the problem, after a detailed analysis of the subtleties of the notion of consequence in the infinite-valued calculus.

4.1 Many-valued propositional calculi

In the early twenties, Jan Łukasiewicz introduced systems of logic in which propositions admit as truth values real numbers between 0 and 1. As main propositional connectives he considered *implication* $\rightarrow$ and *negation* $\neg$, as given by the following "truth tables", where x and y denote arbitrary elements of the real unit interval $[0, 1]$:

(4.1) $x \rightarrow y =_{def} \min(1, 1 - x + y)$

and

(4.2) $\neg x =_{def} 1 - x.$

Along with this infinite-valued propositional system, Łukasiewicz also considered, for each natural number $n \geq 2$, an n-valued system, in which the truth values are the rational numbers

$$0, \frac{1}{n-1}, \ldots, \frac{n-2}{n-1}, 1,$$

and the truth values for implication and negation are again given by formulas (4.1) and (4.2). Thus in particular, for $n = 2$ the only possible truth values are 0 and 1, and the above formulas give back the truth tables for implication and negation in the classical propositional calculus, provided 1 is interpreted as *true* and 0 as *false*.

As the reader will recall, formula (4.2) gives the operation $\neg$ in the MV-algebra $[0, 1]$. On the other hand, formula (4.1) can be written in terms of the MV-operations as follows:

(4.3) $x \rightarrow y = \neg x \oplus y$

whence,

(4.4) $x \oplus y = \neg x \rightarrow y.$

Thus, the operations $\neg$, $\oplus$ and $\rightarrow$ on [0,1] have the same interdefinability properties as negation, disjunction and implication in classical logic. It is natural to consider $\oplus$ as a disjunction connective. Accordingly, equation (1.2) suggests to consider $\odot$ as conjunction.

Given an arbitrary MV-algebra A, let us define the binary operation $\rightarrow$ on A by formula (4.3). That is, for all $x, y \in A$, $x \rightarrow y =_{def} \neg x \oplus y$. Then, trivially, equation (4.4) holds in A.

In the next section we shall consider an equivalent reformulation of MV-algebras in terms of the operations $\rightarrow$, $\neg$ and the constant 1. This will more closely correspond to Łukasiewicz's original presentation of the infinite-valued calculus, which we shall now introduce formally as follows:

Definition 4.1.1 As in the classical case, one starts from the finite alphabet

(4.5) $\quad \Sigma =_{def} \{X, |,), (, \neg, \rightarrow\}.$

(Propositional) formulas are built, exactly as boolean propositions, from a denumerable set of *propositional variables*, $X, X|, X||, \ldots$, for short, $X_0, X_1, X_2, \ldots$, by means of the connectives of *negation* $\neg$ and of *implication* $\rightarrow$.

In more detail, the set **Form** of formulas is given inductively as follows:

F1) Each propositional variable X_k is a formula.

F2) If α is a formula, then $\neg\alpha$ is a formula.

F3) If α and β are formulas, then $(\alpha \rightarrow \beta)$ is a formula.

Notation: For each formula α, **Var**(α) will denote the set of all propositional variables occurring in α.

Definition 4.1.2 Let A be an MV-algebra. Then an *A-valuation* is a function ν: **Form** $\rightarrow A$ satisfying the following properties, where α and β denote arbitrary formulas:

V1) $\quad \nu(\neg\alpha) =_{def} \neg\nu(\alpha)$,

V2) $\nu(\alpha \to \beta) =_{def} \nu(\alpha) \to \nu(\beta)$.

Since unique readability also holds for formulas, any A-valuation ν is uniquely determined by its values $\nu(X_0), \nu(X_1), \ldots$.

An A-valuation ν is said to *A-satisfy a formula* α iff $\nu(\alpha) = 1$; α is an *A-tautology* iff α is A-satisfied by all A-valuations. Formulas α and β are *semantically A-equivalent* iff $\nu(\alpha) = \nu(\beta)$ for all A-valuations ν. From condition (V2) we get that α and β are A-equivalent iff both formulas $\alpha \to \beta$ and $\beta \to \alpha$ are A-tautologies. Let Θ be a set of formulas. We say that a formula α is a *semantic A-consequence of* Θ iff each A-valuation ν that A-satisfies all the formulas in Θ also A-satisfies α. Thus in particular, α is an A-tautology iff α is a semantic A-consequence of the empty set.

Using the identities (4.3), (4.4), every formula α containing the variables $X_1, \ldots, X_k$ can be straightforwardly transformed into an MV-term τ_α in the same variables. Conversely, upon replacing every occurrence of the constant 0 in a term τ by, say, the formula $\neg(X \to X)$, then τ is transformed into a propositional formula α_τ. We shall tacitly identify propositional formulas and MV-terms, using the maps $\alpha \mapsto \tau_\alpha$ and $\tau \mapsto \alpha_\tau$. Also, using all abbreviations introduced in Section 1.4, we shall freely write, e.g., 1 instead of $\neg\neg(X \to X)$.

As a first application of this identification, an easy induction on the number of connectives in α yields

Proposition 4.1.3 *(i) Let A be an MV-algebra and α be a formula, with* $\mathbf{Var}(\alpha) \subseteq \{X_{i_1}, \ldots, X_{i_k}\}$. *Then for each A-valuation ν we have*

$$\nu(\alpha) = \alpha^A(\nu(X_{i_1}), \ldots, \nu(X_{i_k})), \tag{4.6}$$

where $\alpha^A \colon A^k \to A$ is the term function defined in Section 1.4;

(ii) A formula α is an A-tautology if and only if the MV-equation $\alpha = 1$ holds in A;

(iii) Formulas α and β are semantically A-equivalent iff the equation $\alpha = \beta$ holds in A iff $\alpha^A = \beta^A$. □

For each integer $n \geq 2$, the *n-valued Łukasiewicz propositional calculus* deals with propositional formulas equipped with the relation of semantic Ł_n-equivalence. In the *Łukasiewicz infinite-valued propositional calculus* one considers propositional formulas equipped with the relation of semantic [0,1]-equivalence.

By Proposition 4.1.3, the Completeness Theorem (Theorem 2.5.3) has the following equivalent formulation:

Theorem 4.1.4 *A formula α is a $[0,1]$-tautology if, and only if, for every MV-algebra A, α is an A-tautology. Thus, for any two formulas α and β, we have $\alpha^{[0,1]} = \beta^{[0,1]}$ iff $\alpha^A = \beta^A$ for all MV-algebras A.* □

Since our main concern here is the infinite-valued calculus, by a *valuation* we shall henceforth mean a $[0,1]$-valuation; by a *tautology* we shall understand a [0,1]-tautology; also, semantic [0,1]-equivalence and [0,1]-consequence, shall be simply referred to as *semantic equivalence* and *semantic consequence*, respectively.

Notation: For each $\Theta \subseteq \mathbf{Form}$, $\Theta^{\models}$ will denote the set of semantic consequences of Θ. In particular, $\emptyset^{\models}$ will denote the set of all tautologies.

The last two results above allow us to identify term functions $\alpha^{[0,1]}$ and semantic equivalence classes of propositional formulas. Recalling now our analysis of free MV-algebras in Section 3.1, one may reasonably expect that McNaughton functions play in the infinite-valued calculus the same role played by boolean functions in the classical propositional calculus. As an instance of this role, consider the proof of the following result, which will be considerably strengthened in Corollary 4.5.3 and in Theorem 4.6.10 below:

Proposition 4.1.5 *There exists an effective procedure (for definiteness, a Turing machine) enumerating all formulas that are not tautologies.*

Proof: Fixing an arbitrary lexicographic ordering of all strings over alphabet Σ, let $\psi_1, \psi_2, \ldots$ be a list of all n-variable formulas $\psi_t = \psi_t(X_1, \ldots, X_n) \in \mathbf{Form}$. Let similarly $\mathbf{x}^1, \mathbf{x}^2, \ldots$ list all n-tuples $\mathbf{x} = x_1, \ldots, x_n$ of rational numbers in the unit real interval $[0,1]$. Thus, $\mathbf{x}^i = (x_1^i, \ldots, x_n^i)$ for suitable $x_j^i \in [0,1] \cap \mathbf{Q}$. For each $i = 1, 2, \ldots$ let ν^i be the valuation such that $\nu^i(X_j) = \mathbf{x}_j^i$ for all $j = 1, \ldots, n.$, and

$\nu^i(X_m) = 0$ for all remaining variables. Whenever a pair (i, ψ_t) is such that $\nu^i(\psi_t) < 1$, put ψ_t in the list of nontautologies. By Proposition 4.1.3 together with Theorem 4.1.4 and our analysis of free MV-algebras in Section 3, the desired conclusion now follows from the continuity of McNaughton functions: indeed, if a formula ψ is not a tautology then there is a valuation ν whose values are all rational, such that $\nu(\psi) < 1$. □

In the next sections we shall obtain an effective procedure to enumerate all tautologies.

4.2 Wajsberg algebras

Stone's representation theorem for boolean algebras allows one to visualize the connectives of negation, disjunction and conjunction of the classical propositional calculus as the set-theoretical operations of complement, union and intersection. In the infinite-valued case, since $\oplus$ and $\odot$, together with negation $\neg$ can express $\vee$ and $\wedge$, the (additive) operations $\oplus$ and $\odot$ are regarded as more fundamental than the (lattice) operations $\vee$ and $\wedge$.

Historically, Chang introduced MV-algebras as generalized boolean algebras; indeed, Chang's completeness theorem is a generalization of the fact that if an equation holds in the boolean algebra $\{0, 1\}$, then the equation holds in every boolean algebra. Already the proof of the one-variable case of McNaughton's theorem given in Section 3 enhances the role of the associative-commutative operation $\oplus$, giving normal form reductions of all elements of $Free_1$ as *sums* of Schauder hat functions. In a later chapter we shall establish a categorical equivalence between MV-algebras and abelian ℓ-groups with strong unit (the latter being deeply related to the time-honoured theory of magnitudes). Altogether, the operations $\odot$ and $\oplus$ have a distinctly arithmetic, rather than a set-theoretic significance.

On the other hand, the identities $x \to y = \neg x \oplus y$ and $x \oplus y = \neg x \to y$ show that there is no essential difference of expressive power between the additive connectives and the implication connective. It turns out that for our analysis of proofs and consequence in this chapter, it is more convenient to replace the $\odot$ and $\oplus$ operations by the implication connective $\rightarrow$. In this way one obtains the following equivalent variant of MV-algebras:

Definition 4.2.1 A *Wajsberg* algebra (for short, a *W*-algebra) is a system $A = \langle A, \rightarrow, \neg, 1\rangle$, where A is a nonempty set, and the binary operation $\rightarrow$, the unary operation $\neg$ and the distinguished element 1 satisfy the following equations:

W1) $1 \rightarrow x = x$

W2) $(x \rightarrow y) \rightarrow ((y \rightarrow z) \rightarrow (x \rightarrow z)) = 1$

W3) $(x \rightarrow y) \rightarrow y = (y \rightarrow x) \rightarrow x$

W4) $(\neg x \rightarrow \neg y) \rightarrow (y \rightarrow x) = 1$.

Lemma 4.2.2 *Let A be an MV-algebra, and put $x \rightarrow y =_{def} \neg x \oplus y$ and $1 =_{def} \neg 0$. Then $\langle A, \rightarrow, \neg, 1\rangle$ is a W-algebra.* □

Our next task is to prove that conversely, each W-algebra becomes an MV-algebra, once equipped with the operations $\neg$, $x \oplus y =_{def} \neg x \rightarrow y$ and $0 =_{def} \neg 1$.

Lemma 4.2.3 *Let $A = \langle A, \rightarrow, 1\rangle$ be a system satisfying (W1), (W2) and (W3). Then the following properties hold for every x, y and z in A:*

W5) $x \rightarrow x = 1$

W6) *If* $x \rightarrow y = y \rightarrow x = 1$, *then* $x = y$

W7) $x \rightarrow 1 = 1$

W8) $x \rightarrow (y \rightarrow x) = 1$

W9) *If* $x \rightarrow y = y \rightarrow z = 1$, *then* $x \rightarrow z = 1$

W10) $(x \rightarrow y) \rightarrow ((z \rightarrow x) \rightarrow (z \rightarrow y)) = 1$

W11) $x \rightarrow (y \rightarrow z) = y \rightarrow (x \rightarrow z)$.

Proof: W5) By (W2), $(1 \to 1) \to ((1 \to x) \to (1 \to x)) = 1$, and then, by (W1), $x \to x = 1 \to (x \to x) = 1$.

W6) If $x \to y = y \to x = 1$, then by (W1) and (W3) we have $x = 1 \to x = (y \to x) \to x = (x \to y) \to y = 1 \to y = y$.

W7) By (W3), (W1) and (W5), $(x \to 1) \to 1 = (1 \to x) \to x = x \to x = 1$. From this identity, (W2), (W1) and (W5) we have $1 = (1 \to x) \to ((x \to 1) \to (1 \to 1)) = x \to ((x \to 1) \to 1) = x \to 1$.

W8) By (W2), (W7) and (W1), $1 = (y \to 1) \to ((1 \to x) \to (y \to x)) = 1 \to (x \to (y \to x)) = x \to (y \to x)$.

W9) If $x \to y = y \to z = 1$, then by (W2), $1 = (x \to y) \to ((y \to z) \to (x \to z)) = 1 \to (1 \to (x \to z))$, and applying (W1) twice, we obtain $x \to z = 1$.

W10) As a preliminary step, we shall prove the following weaker form of (W11):

W11′) If $x \to (y \to z) = 1$, then $y \to (x \to z) = 1$.

Suppose that $x \to (y \to z) = 1$. Replacing y by $y \to z$ in (W2) we get $1 = (x \to (y \to z)) \to (((y \to z) \to z) \to (x \to z)) = 1 \to (((y \to z) \to z) \to (x \to z))$. Hence, by (W1) and (W3), $((z \to y) \to y) \to (x \to z) = 1$. On the other hand, by (W8), $y \to ((z \to y) \to y) = 1$. Applying (W9) to these identities, with $x = y$, $y = (z \to y) \to y$ and $z = x \to z$, we obtain $y \to (x \to z) = 1$. Now (W10) follows at once from (W11′) and (W2).

W11) By (W3) and (W8), $y \to ((y \to z) \to z) = y \to ((z \to y) \to y) = 1$, and by W10), $((y \to z) \to z) \to ((x \to (y \to z)) \to (x \to z)) = 1$. Applying (W9) to the last two identities, we obtain $y \to ((x \to (y \to z)) \to (x \to z)) = 1$. Hence, by (W11′), $(x \to (y \to z)) \to (y \to (x \to z)) = 1$. By interchanging x and y in this identity we also obtain $(y \to (x \to z)) \to (x \to (y \to z)) = 1$, and then we can apply (W6) to obtain (W11). □

Lemma 4.2.4 *The following equations hold in every W-algebra A:*

(i) $\neg 1 \to x = 1$;

(ii) $\neg x = x \to \neg 1$;

(iii) $\neg\neg x = x$;

(iv) $x \to y = \neg y \to \neg x$.

Proof: (i) By (W10) we have $((\neg x \to \neg 1) \to x) \to ((\neg 1 \to (\neg x \to \neg 1) \to (\neg 1 \to x)) = 1$. Since, by (W1) and (W4), $(\neg x \to \neg 1) \to x = (\neg x \to \neg 1) \to (1 \to x) = 1$ and, by (W8), $\neg 1 \to (\neg x \to \neg 1) = 1$, whence the desired conclusion follows from (W1).

(ii) By (W8), $\neg x \to (\neg\neg 1 \to \neg x) = 1$. By (W4), $(\neg\neg 1 \to \neg x) \to (x \to \neg 1) = 1$. Then from (W9) we obtain $\neg x \to (x \to \neg 1) = 1$, and applying (W11) we have $x \to (\neg x \to \neg 1) = 1$. As seen in the proof of (i), we also have $(\neg x \to \neg 1) \to x = 1$. Hence by (W6) we obtain $x = \neg x \to \neg 1$, and taking into account (W3), together with (i) and (W1) we get $x \to \neg 1 = (\neg x \to \neg 1) \to \neg 1 = (\neg 1 \to \neg x) \to \neg x = 1 \to \neg x = \neg x$.

(iii) By (ii), (W3), (i) and (W1) we can write $\neg\neg x = (x \to \neg 1) \to \neg 1 = (\neg 1 \to x) \to x = 1 \to x = x$.

(iv) By (W4) and (iii) we have $1 = (\neg\neg x \to \neg\neg y) \to (\neg y \to \neg x) = (x \to y) \to (\neg y \to \neg x)$. Now apply (W4) and (W6). □

Theorem 4.2.5 *Let $\langle A, \to, \neg, 1\rangle$ be a Wajsberg algebra. Upon defining $x \oplus y =_{def} \neg x \to y$ and $0 =_{def} \neg 1$, the system $\langle A, \oplus, \neg, 0\rangle$ is an MV-algebra.*

Proof: $\langle A, \oplus, \neg, 0\rangle$ satisfies MV1), because

$$x \oplus y = \neg x \to y = \neg y \to \neg\neg x = y \oplus x,$$

and also satisfies MV2), as shown by the following identities:

$$x \oplus (y \oplus z) = \neg x \to (\neg y \to z) = \neg x \to (\neg z \to y)$$

$$= \neg z \to (\neg x \to y) = z \oplus (x \oplus y) = (x \oplus y) \oplus z.$$

The remaining verifications, MV3)–MV6), are easy consequences of the properties established in the previous lemmas. □

In what follows, $\rightarrow$ will denote the binary operation defined on MV-algebras by formula (4.3). In the light of the above theorem, this operation satisfies properties (W1) - (W11).

Lemma 1.1.4(iii) gives the following remarkable relation between the operations $\rightarrow$ and $\odot$ on an MV-algebra A:

(4.7) For all $x, y, z \in A$, $z \odot x \leq y$ iff $z \leq x \rightarrow y$.

Definition 4.2.6 An *implicative filter* of an MV-algebra A is a subset F of A satisfying the following conditions:

(F1) $1 \in F$;

(F2) For all x, y in A, if $x \in F$ and $x \rightarrow y \in F$, then $y \in F$.

An implicative filter F of A is said to be *proper* iff $F \neq A$; F is called *maximal* iff F is proper and A is the only implicative filter strictly containing F.

Lemma 4.2.7 *The following are equivalent conditions for each subset F of an MV-algebra A:*

(i) *F is an implicative filter;*

(ii) *$F \neq \emptyset$; if $x \in F$ and $x \leq y \in A$ then $y \in F$; if $x, y \in F$ then $x \odot y \in F$;*

(iii) *The set $\neg F =_{def} \{\neg x \mid x \in F\}$ is an ideal of A.*

Proof: (i) implies (ii): Let F be an implicative filter. By (F1), $1 \in F$. Suppose $x \in F$ and $x \leq y$. Since this last condition is equivalent to $x \rightarrow y = 1$, (F2) yields $y \in F$. Note that $y \rightarrow (x \rightarrow (x \odot y)) = 1$. Therefore if x, y are both in F, then by (F2) we obtain that $x \odot y \in F$.

(ii) implies (iii): This immediately follows by definition of the operation $\odot$ as given by equation (1.2).

(iii) implies (i): Suppose that $\neg F$ is an ideal. Since $0 \in \neg F$, we have $1 \in F$. Suppose now that x and $x \rightarrow y$ are in F. This means that $\neg x \in \neg F$ and $x \ominus y \in \neg F$, and then $\neg x \vee \neg y = \neg x \oplus (x \ominus y) \in \neg F$. Therefore, $\neg y \in \neg F$, i.e., $y \in F$. □

The above lemma allows us to obtain properties of implicative filters from properties of ideals. For instance, if $h: A \to A'$ is a homomorphism then the set $F(h) =_{def} \{x \in A \mid h(x) = 1\}$ is an implicative filter; moreover, $h(x) = h(y)$ iff $(x \to y) \wedge (y \to x) \in F(h)$.

Notation: Given an implicative filter F, by abuse of notation we shall write A/F to denote the quotient algebra $A/\neg F$.

The intersection of any nonempty family of implicative filters of an MV-algebra A is an implicative filter, and since A itself is an implicative filter, we can give the following

Definition 4.2.8 For any subset X of an MV-algebra A, the *implicative filter $\mathcal{D}(X)$ generated by X* is the intersection of all implicative filters containing X.

Proposition 4.2.9 *Let A be an arbitrary MV-algebra.*

(i) Let $s, t \in A$ and $Y \subseteq A$. Then $t \in \mathcal{D}(Y \cup \{s\})$ iff there is an integer $n \geq 1$ such that $s^n \to t \in \mathcal{D}(Y)$;

(ii) For any $s, t \in A$ we have $t \in \mathcal{D}(\{s\})$ iff there is an integer $n \geq 1$ such that $s^n \to t = 1$;

(iii) For arbitrary $x_1, \ldots, x_k \in A$ we have

$$\mathcal{D}(\{x_1, \ldots, x_k\}) = \mathcal{D}(x_1 \odot \ldots \odot x_k). \tag{4.8}$$

Proof: (i) and (ii) can be easily derived from (1.8) and (1.9), respectively. To prove (iii), let $n_1, \ldots, n_k \geq 1$ be integers, and $n = \max(n_1, \ldots, n_k)$. Then the desired conclusion follows from the inequality $(x_1 \odot \ldots \odot x_k)^n \leq x_1^{n_1} \odot \ldots \odot x_k^{n_k}$. □

4.3 Provability

Definition 4.3.1 An *axiom* of the Łukasiewicz infinite-valued propositional calculus is a formula that can be written in any one of the following ways, where α, β and γ denote arbitrary formulas:

(A1) $\alpha \to (\beta \to \alpha)$;

(A2) $(\alpha \to \beta) \to ((\beta \to \gamma) \to (\alpha \to \gamma))$;

(A3) $((\alpha \to \beta) \to \beta) \to ((\beta \to \alpha) \to \alpha)$;

(A4) $(\neg\alpha \to \neg\beta) \to (\beta \to \alpha)$.

Note that one can effectively decide whether a given string of symbols over the alphabet Σ of (4.5) is an axiom.

Definition 4.3.2 A *proof from a set* Θ *of formulas* is a finite string of formulas $\alpha_1, \ldots, \alpha_n$, with $n \geq 1$, such that, for each $1 \leq i \leq n$:

(i) α_i is an axiom, or

(ii) $\alpha_i \in \Theta$, or

(iii) there are $j, k \in \{1, \ldots, i-1\}$ such that α_k coincides with the formula $(\alpha_j \to \alpha_i)$.

A formula α is *provable from* Θ, in symbols, $\Theta \vdash \alpha$, iff there is a proof $\alpha_1, \ldots, \alpha_n$ from Θ, such that $\alpha_n = \alpha$. Equivalently, we say that α is a *syntactic consequence of* Θ. The set of provable formulas from Θ shall be denoted

$$\Theta^{\vdash}.$$

As usual, we say that formula β *follows by modus ponens from formulas* α *and* $\alpha \to \beta$. With this terminology, condition (iii) in the above definition can be given by the following equivalent reformulation:

(iii′) there are $j, k \in \{1, \ldots, i-1\}$ such that α_i follows by modus ponens from α_j and α_k.

In Definition 4.3.2 the case $\Theta = \emptyset$ is not excluded. In this case, condition (ii) becomes vacuous.

Definition 4.3.3 By a *proof* we shall henceforth mean a proof from the empty set; by a *provable* formula we shall mean a formula that is provable from the empty set. We shall also write $\vdash \alpha$ in place of $\emptyset \vdash \alpha$.

Given a finite sequence of strings of symbols from the alphabet Σ we can effectively decide whether such string is a proof. Using some lexicographic ordering, we then see that *there exists an effective procedure to list the set of provable formulas.*

Our next aim is to show that *the tautologies coincide with provable formulas*; this, together with our discussion in the first sections, will give a positive solution to the decision problem for the infinite-valued calculus of Łukasiewicz.

Notation: We shall use $\alpha \vee \beta$ and $\alpha \wedge \beta$ as abbreviations of $(\alpha \to \beta) \to \beta$ and $\neg((\neg\alpha \to \neg\beta) \to \neg\beta)$, respectively. With this notation, (A3) can be written as $(\alpha \vee \beta) \to (\beta \vee \alpha)$.

Proposition 4.3.4 *For all formulas α, β and γ we have*

(4.9) $\vdash (\alpha \to (\beta \to \gamma)) \to (\beta \to (\alpha \to \gamma))$

(4.10) $\vdash \alpha \to \alpha$

(4.11) $\vdash (\beta \to \gamma) \to ((\alpha \to \beta) \to (\alpha \to \gamma))$

(4.12) $\vdash \neg\neg\alpha \to (\beta \to \alpha)$

(4.13) $\vdash \neg\neg\alpha \to \alpha$

(4.14) $\vdash (\alpha \to \neg\beta) \to (\beta \to \neg\alpha)$

(4.15) $\vdash (\alpha \to \neg\neg\beta) \to (\neg\beta \to \neg\alpha)$

(4.16) $\vdash \alpha \to \neg\neg\alpha$.

Proof: To prove (4.9) it suffices to check that the following string of formulas is a proof. To help the reader, on the right-hand side of each formula occurring in the string, we shall point out whether this formula is an axiom, or else the formula follows by modus ponens from preceding formulas.

$\alpha_1 : \beta \to (\gamma \vee \beta) = \beta \to ((\gamma \to \beta) \to \beta)$ (A1)

$\alpha_2 : (\beta \to (\gamma \vee \beta)) \to (((\gamma \vee \beta) \to (\beta \vee \gamma)) \to (\beta \to (\beta \vee \gamma)))$ (A2)

$\alpha_3 : ((\gamma \vee \beta) \rightarrow (\beta \vee \gamma)) \rightarrow (\beta \rightarrow (\beta \vee \gamma))$ (MP-α_1, α_2)

$\alpha_4 : (\gamma \vee \beta) \rightarrow (\beta \vee \gamma)$ (A3)

$\alpha_5 : \beta \rightarrow (\beta \vee \gamma)$ (MP-α_4, α_3)

$\alpha_6 : (\beta \rightarrow (\beta \vee \gamma)) \rightarrow (((\beta \vee \gamma) \rightarrow (\alpha \rightarrow \gamma)) \rightarrow (\beta \rightarrow (\alpha \rightarrow \gamma)))$ (A2)

$\alpha_7 : ((\beta \vee \gamma) \rightarrow (\alpha \rightarrow \gamma)) \rightarrow (\beta \rightarrow (\alpha \rightarrow \gamma))$ (MP-α_5, α_6)

$\alpha_8 : (\alpha \rightarrow (\beta \rightarrow \gamma)) \rightarrow (((\beta \rightarrow \gamma) \rightarrow \gamma) \rightarrow (\alpha \rightarrow \gamma))$
$= (\alpha \rightarrow (\beta \rightarrow \gamma)) \rightarrow ((\beta \vee \gamma) \rightarrow (\alpha \rightarrow \gamma))$ (A2)

$\alpha_9 : \alpha_8 \rightarrow (\alpha_7 \rightarrow ((\alpha \rightarrow (\beta \rightarrow \gamma)) \rightarrow (\beta \rightarrow (\alpha \rightarrow \gamma))))$ (A2)

$\alpha_{10} : \alpha_7 \rightarrow ((\alpha \rightarrow (\beta \rightarrow \gamma)) \rightarrow (\beta \rightarrow (\alpha \rightarrow \gamma)))$ (MP-α_8, α_9)

$\alpha_{11} : (\alpha \rightarrow (\beta \rightarrow \gamma)) \rightarrow (\beta \rightarrow (\alpha \rightarrow \gamma))$ (MP-α_7, α_{10})

This settles (4.9). To prove (4.10) note first that, taking $\gamma = \alpha$ in (4.9), we obtain

$$\vdash (\alpha \rightarrow (\beta \rightarrow \alpha)) \rightarrow (\beta \rightarrow (\alpha \rightarrow \alpha)).$$

If $\alpha_1, \ldots, \alpha_n = (\alpha \rightarrow (\beta \rightarrow \alpha)) \rightarrow (\beta \rightarrow (\alpha \rightarrow \alpha))$ is a proof, then

$$\alpha_1, \ldots, \alpha_n,\ \alpha_{n+1} = \alpha \rightarrow (\beta \rightarrow \alpha),\ \alpha_{n+2} = \beta \rightarrow (\alpha \rightarrow \alpha)$$

is also a proof. Hence taking any axiom as β, we obtain that the sequence $\alpha_1, \ldots, \alpha_n, \alpha_{n+1}, \alpha_{n+2}, \alpha_{n+3} = \beta, \alpha_{n+4} = \alpha \rightarrow \alpha$ is a proof of $\alpha \rightarrow \alpha$.

To prove (4.11), as a particular case of (4.9) we have

$$\vdash \delta = ((\alpha \rightarrow \beta) \rightarrow ((\beta \rightarrow \gamma) \rightarrow (\alpha \rightarrow \gamma))) \rightarrow$$
$$((\beta \rightarrow \gamma) \rightarrow ((\alpha \rightarrow \beta) \rightarrow (\alpha \rightarrow \gamma))).$$

If $\alpha_1, \ldots, \alpha_n = \delta$ is a proof, then the string

$\alpha_1, \ldots, \alpha_n,$

$\alpha_{n+1} = (\alpha \rightarrow \beta) \rightarrow ((\beta \rightarrow \gamma) \rightarrow (\alpha \rightarrow \gamma)),$

$$\alpha_{n+2} = (\beta \to \gamma) \to ((\alpha \to \beta) \to (\alpha \to \gamma))$$

is also a proof. This settles (4.11).

To prove (4.12) it is enough to check that the following string of formulas is a proof:

$$\alpha_1 : \neg\neg\alpha \to (\neg\neg\beta \to \neg\neg\alpha) \qquad \text{(A1)}$$

$$\alpha_2 : (\neg\neg\beta \to \neg\neg\alpha) \to (\neg\alpha \to \neg\beta) \qquad \text{(A4)}$$

$$\alpha_3 : (\neg\neg\alpha \to (\neg\neg\beta \to \neg\neg\alpha)) \to (((\neg\neg\beta \to \neg\neg\alpha) \to (\neg\alpha \to \neg\beta)) \to (\neg\neg\alpha \to (\neg\alpha \to \neg\beta))) \qquad \text{(A2)}$$

$$\alpha_4 : ((\neg\neg\beta \to \neg\neg\alpha) \to (\neg\alpha \to \neg\beta)) \to (\neg\neg\alpha \to (\neg\alpha \to \neg\beta)) \qquad \text{(MP-}\alpha_1, \alpha_3)$$

$$\alpha_5 : \neg\neg\alpha \to (\neg\alpha \to \neg\beta) \qquad \text{(MP- }\alpha_2, \alpha_4)$$

$$\alpha_6 : (\neg\alpha \to \neg\beta) \to (\beta \to \alpha) \qquad \text{(A4)}$$

$$\alpha_7 : (\neg\neg\alpha \to (\neg\alpha \to \neg\beta)) \to (((\neg\alpha \to \neg\beta) \to (\beta \to \alpha)) \to (\neg\neg\alpha \to (\beta \to \alpha))) \qquad \text{(A2)}$$

$$\alpha_8 : ((\neg\alpha \to \neg\beta) \to (\beta \to \alpha)) \to (\neg\neg\alpha \to (\beta \to \alpha)) \qquad \text{(MP- }\alpha_5, \alpha_7)$$

$$\alpha_9 : \neg\neg\alpha \to (\beta \to \alpha) \qquad \text{(MP- }\alpha_6, \alpha_8).$$

This settles (4.12).

To prove (4.13), combining (4.9) and (4.12) we easily obtain $\vdash \beta \to (\neg\neg\alpha \to \alpha)$. Since β is arbitrary, letting β be an axiom we immediately get the desired proof.

To prove (4.14) it is enough to verify that a new proof is obtained upon adding to any proof $\alpha_1, \ldots, \alpha_n$ of $\neg\neg\alpha \to \alpha$ the following formulas:

$$\alpha_{n+1} : (\neg\neg\alpha \to \alpha) \to ((\alpha \to \neg\beta) \to (\neg\neg\alpha \to \neg\beta)) \qquad \text{(A2)}$$

$$\alpha_{n+2} : (\alpha \to \neg\beta) \to (\neg\neg\alpha \to \neg\beta) \qquad \text{(MP-}\alpha_n, \alpha_{n+1})$$

$$\alpha_{n+3} : ((\alpha \to \neg\beta) \to (\neg\neg\alpha \to \neg\beta)) \to (((\neg\neg\alpha \to \neg\beta) \to$$

$$(\beta \to \neg\alpha)) \to ((\alpha \to \neg\beta) \to (\beta \to \neg\alpha))) \qquad \text{(A2)}$$

$\alpha_{n+4} : ((\neg\neg\alpha \to \neg\beta) \to (\beta \to \neg\alpha)) \to$
$((\alpha \to \neg\beta) \to (\beta \to \neg\alpha))$ (MP-$\alpha_{n+2}, \alpha_{n+3}$)

$\alpha_{n+5} : (\neg\neg\alpha \to \neg\beta) \to (\beta \to \neg\alpha)$ (A4)

$\alpha_{n+6} : (\alpha \to \neg\beta) \to (\beta \to \neg\alpha)$ (MP- $\alpha_{n+5}, \alpha_{n+4}$).

To obtain (4.15) it is sufficient to replace β by $\neg\beta$ in (4.14).

Finally, to prove (4.16), replacing α in (4.14) by $\neg\alpha$ and β by α, we get $\vdash (\neg\alpha \to \neg\alpha) \to (\alpha \to \neg\neg\alpha)$. Combining now this result with (4.10), in which α is replaced by $\neg\alpha$, we get the desired conclusion. □

From (A1) and modus ponens we immediately obtain

(4.17) If $\vdash \alpha$ then, for each β, $\vdash \beta \to \alpha$.

4.4 Lindenbaum algebra

Theorem 4.4.1 *Let the binary relation $\equiv$ on* **Form** *be defined by $\alpha \equiv \beta$ iff $\vdash \alpha \to \beta$ and $\vdash \beta \to \alpha$. Then $\equiv$ is an equivalence relation, called* syntactic equivalence , *satisfying the following conditions:*

(4.18) *If $\alpha \equiv \gamma$ and $\beta \equiv \delta$ then $(\alpha \to \beta) \equiv (\gamma \to \delta)$;*

(4.19) *If $\alpha \equiv \beta$ then $\neg\alpha \equiv \neg\beta$.*

Proof: It is obvious that $\alpha \equiv \beta$ implies $\beta \equiv \alpha$ and, by (4.10), $\alpha \equiv \alpha$. By (A2), $\{\alpha \to \beta, \beta \to \gamma\} \vdash \alpha \to \gamma$ and $\{\gamma \to \beta, \beta \to \alpha\} \vdash \gamma \to \alpha$. Hence, $\equiv$ is transitive, and we have shown that $\equiv$ is an equivalence relation on **Form**. By (A2), for any formulas α, β and γ we have

$$\{\gamma \to \alpha\} \vdash (\alpha \to \beta) \to (\gamma \to \beta)$$

and

$$\{\alpha \to \gamma\} \vdash (\gamma \to \beta) \to (\alpha \to \beta).$$

Therefore,

(4.20) If $\alpha \equiv \gamma$ then $\alpha \to \beta \equiv \gamma \to \beta$.

Analogously, from (4.11) we obtain

(4.21) If $\beta \equiv \delta$ then $\gamma \to \beta \equiv \gamma \to \delta$.

Now (4.18) follows from (4.20), (4.21) and the transitivity of $\equiv$.

By (4.13) and (4.16), for every formula β, $\beta \equiv \neg\neg\beta$. Hence by (4.21), for arbitrary formulas α and β, $\alpha \to \beta \equiv \alpha \to \neg\neg\beta$. Therefore, $\vdash (\alpha \to \beta) \to (\alpha \to \neg\neg\beta)$. From this result and (4.15) we obtain $\vdash (\alpha \to \beta) \to (\neg\beta \to \neg\alpha)$. Since by (A4), $\vdash (\neg\beta \to \neg\alpha) \to (\alpha \to \beta)$, we have (4.19). □

Notation: The equivalence class of formula α with respect to syntactic equivalence shall be denoted by $|\alpha|$. In symbols,

$$|\alpha| =_{def} \{\beta \in \mathbf{Form} \mid \beta \equiv \alpha\}.$$

Lemma 4.4.2 *For each formula α, $\alpha \in \emptyset^{\vdash}$ iff $|\alpha| = \emptyset^{\vdash}$.*

Proof: If $|\alpha| = \emptyset^{\vdash}$ then $\alpha \in \emptyset^{\vdash}$ because $\alpha \in |\alpha|$. Conversely, suppose $\alpha \in \emptyset^{\vdash}$, i.e., $\vdash \alpha$. Take $\beta \in \emptyset^{\vdash}$. By (4.17), $\beta \equiv \alpha$. Hence $\emptyset^{\vdash} \subseteq |\alpha|$. Now take $\beta \in |\alpha|$. Then $\vdash (\alpha \to \beta)$ and, by modus ponens, $\vdash \beta$. Hence $|\alpha| \subseteq \emptyset^{\vdash}$. □

Theorem 4.4.3 *The quotient set* **Form**/$\equiv$ *becomes a Wajsberg algebra, once equipped with the operations $\to$ and $\neg$ and the constant* 1 *as given by the following stipulations:*

(4.22) $|\alpha| \to |\beta| =_{def} |\alpha \to \beta|$

(4.23) $\neg|\alpha| =_{def} |\neg\alpha|$

(4.24) $1 =_{def} \emptyset^{\vdash}$.

Proof: By (4.18) and (4.19) in Theorem 4.4.1, it follows that the two identities (4.23) and (4.24) yield well defined operations on the quotient set **Form**/$\equiv$. Moreover, by Lemma 4.4.2, $\emptyset^{\vdash} \in$ **Form**/$\equiv$. Then it remains to verify that the operations defined by (4.22) and (4.23) and the constant 1 as defined by (4.24) satisfy equations (W1) - (W4) in Definition 4.2.1. To prove (W1), note first that $1 \to |\alpha| = \emptyset^{\vdash} \to |\alpha| =$

$|\beta \to \alpha|$, where $\beta \in \emptyset^{\vdash}$. By (A1) we have $\vdash \alpha \to (\beta \to \alpha)$. On the other hand, by (A3), $(\beta \to \alpha) \to \alpha \equiv (\alpha \to \beta) \to \beta$ and, since $\vdash \beta$, it follows from (4.17) that $\vdash (\alpha \to \beta) \to \beta$. Hence, by Lemma 4.4.2 we have $\vdash (\beta \to \alpha) \to \alpha$, and then $\beta \to \alpha \equiv \alpha$. Consequently, $1 \to |\alpha| = \emptyset^{\vdash} \to |\alpha| = |\beta \to \alpha| = |\alpha|$, and this proves (W1). In the light of Lemma 4.4.2, conditions (W2), (W3) and (W4) follow at once from (A2), (A3) and (A4), respectively. □

By Theorem 4.2.5 we get

Corollary 4.4.4 *The quotient set* **Form**$/\equiv$ *becomes an MV-algebra with the operations* $\neg$ *and* $\oplus$ *and the constant* 0 *defined by*

(4.25) $\neg|\alpha| =_{def} |\neg\alpha|$

(4.26) $|\alpha| \oplus |\beta| =_{def} |\neg\alpha \to \beta|$

(4.27) $0 =_{def} \neg\emptyset^{\vdash} = \{\alpha \in \mathbf{Form} \mid$ *there is* $\beta \in \emptyset^{\vdash}$ *such that* $\alpha \equiv \neg\beta\}$.

□

The MV-algebra

$$\mathcal{L} =_{def} \langle \mathbf{Form}/\equiv, 0, \neg, \oplus \rangle$$

is called the *Lindenbaum algebra of Łukasiewicz infinite-valued propositional calculus.*

4.5 All tautologies are provable

Lemma 4.5.1 *Each formula provable from a set* Θ *of formulas is also a semantic consequence of this set. In symbols,*

$$\Theta^{\vdash} \subseteq \Theta^{\models}.$$

Proof: Let ν: **Form** $\to [0,1]$ be a valuation such that $\nu(\alpha) = 1$ for all $\alpha \in \Theta$. By induction on n we shall prove that if $\alpha_1, \ldots, \alpha_n$ is a proof from Θ, then $\nu(\alpha_n) = 1$.

If $n = 1$, then α_1 is either an axiom or it belongs to Θ. In the second case, $\nu(\alpha_1) = 1$ by the hypothesis on ν. In the first case we also have $\nu(\alpha_1) = 1$ because all axioms are tautologies, as shown by direct inspection of (A1) - (A4).

Let $n > 1$ and suppose that, for each proof from Θ, $\beta_1, \ldots, \beta_m$, with $m < n$, we have $\nu(\beta_m) = 1$ (induction hypothesis). Let $\alpha_1, \ldots, \alpha_n$ be a proof from Θ. If α_n is not an axiom and does not belong to Θ, then there are $i, j \in \{1, \ldots, n\}$ such that α_j coincides with the formula $(\alpha_i \to \alpha_n)$. Since both $\alpha_1, \ldots, \alpha_i$ and $\alpha_1, \ldots, \alpha_j$ are proofs from Θ, by induction hypothesis, $\nu(\alpha_i) = \nu(\alpha_j) = 1$. Therefore,

$$1 = \nu(\alpha_j) = 1 \to \nu(\alpha_n) = \nu(\alpha_n).$$

□

Thus, in particular, *all provable formulas are tautologies.* The converse follows from Chang's completeness theorem:

Theorem 4.5.2 *Every tautology is provable. Thus, tautologies coincide with provable formulas, in symbols,* $\emptyset^{\models} = \emptyset^{\vdash}$.

Proof: First of all, for each propositional variable X_i, the syntactic equivalence class $|X_i|$ is an element of the Lindenbaum algebra $\mathcal{L}$. Let α be a formula with $\mathbf{Var}(\alpha) \subseteq \{X_{i_1}, \ldots, X_{i_n}\}$. By an easy induction on the number of connectives in α, it follows that $\alpha^{\mathcal{L}}(|X_{i_1}|, \ldots, |X_{i_n}|) = |\alpha|$. Thus, in case $\alpha \in \mathbf{Form}$ is not provable, then by Lemma 4.4.2 and (4.24), $|\alpha| \neq 1$, whence $\alpha^{\mathcal{L}}(|X_{i_1}|, \ldots, |X_{i_n}|) \neq 1$. Stated otherwise, the Lindenbaum algebra $\mathcal{L}$ does not satisfy the equation $\alpha = 1$. Hence by Chang's completeness theorem 2.5.3, the MV-algebra [0,1] does not satisfy this equation, i.e., α is not a tautology. The rest follows from Lemma 4.5.1. □

Corollary 4.5.3 *There is an effective procedure (for definiteness, a Turing machine) deciding whether an arbitrary formula is a tautology in the infinite-valued calculus of Łukasiewicz.*

Proof: By the above theorem, together with the remark following Definition 4.3.3, tautologies can be effectively enumerated. The effective enumerability of nontautologies was shown in Proposition 4.1.5. □

This result shall be considerably strengthened in a later chapter, where we shall give a more accurate estimate of the number of Turing steps needed to decide whether a formula is a tautology.

Corollary 4.5.4 *The relations of semantic and syntactic equivalence coincide: thus, for arbitrary formulas* α *and* β, $\alpha^{[0,1]} = \beta^{[0,1]}$ *iff* $\alpha \equiv \beta$. □

Following tradition, whenever two formulas are syntactically (= semantically) equivalent, we shall henceforth say that they are *logically equivalent.*

Proposition 4.5.5 *Up to isomorphism, the Lindenbaum algebra* $\mathcal{L}$ *coincides with the free MV-algebra over the generating set* $\{|X_0|, |X_1|, \ldots\}$ *of logical equivalence classes of propositional variables.*

Proof: Recalling the terminology introduced in Section 3.1, let us define the map $\varphi: \mathcal{L} \to Term([0,1], \omega)$ by the stipulation $\varphi(|\alpha|) = \alpha^{[0,1]}$. Then Corollary 4.5.4 implies that φ is an isomorphism of the Lindenbaum algebra $\mathcal{L}$ onto the term algebra $Term([0,1], \omega)$. In particular, the restriction of φ to the set of logical equivalence classes of propositional variables yields a bijection from this set onto the set of projection functions $\{\pi_0, \pi_1, \ldots\}$. Hence, the desired conclusion follows from Proposition 3.1.4. □

Lemma 4.5.6 *Let* A *be an MV-algebra and* ν: **Form** $\to A$ *be an A-valuation. Then the stipulation*

$$h_\nu(|\alpha|) = \nu(\alpha)$$

defines a homomorphism $h_\nu: \mathcal{L} \to A$. *Conversely, for each homomorphism* $h: \mathcal{L} \to A$, *the stipulation* $\nu_h(\alpha) = h(|\alpha|)$ *defines an A-valuation* ν_A: **Form** $\to A$. *Moreover, the correspondence* $\nu \mapsto h_\nu$ *is a one-one mapping from the set of A-valuations onto the set of homomorphisms into* A *of the Lindenbaum algebra* $\mathcal{L}$. *The inverse mapping is given by* $h \mapsto \nu_h$.

Proof: Since $\beta \in |\alpha|$ implies $\nu(\beta) = \nu(\alpha)$, the stipulation $h_\nu(|\alpha|) = \nu(\alpha)$ defines a mapping $h_\nu: \mathcal{L} \to A$. The rest of the proof is obvious. □

4.6 Syntactic and semantic consequence

By Lemma 4.5.1, for any set Θ of formulas we have the inclusion $\Theta^{\vdash} \subseteq \Theta^{\models}$. By Theorem 4.5.2, $\emptyset^{\models} = \emptyset^{\vdash}$. As we shall see in this section, in general, $\Theta^{\models} \neq \Theta^{\vdash}$; we shall give necessary and sufficient conditions for $\Theta^{\models}$ to coincide with $\Theta^{\vdash}$.

Definition 4.6.1 A *theory* of Łukasiewicz infinite-valued propositional calculus is a set Θ of formulas satisfying the following conditions:

(T1) All axioms belong to Θ;

(T2) If $\alpha \in \Theta$ and $(\alpha \to \beta) \in \Theta$, then $\beta \in \Theta$.

Proposition 4.6.2 *The following conditions hold for each set Θ of formulas:*

(i) $\Theta^{\vdash}$ is the smallest theory containing Θ;

(ii) Θ is a theory iff $\Theta = \Theta^{\vdash}$;

(iii) If Θ is a theory and $\alpha \in \Theta$, then $|\alpha| \subseteq \Theta$.

Proof: (i) It follows at once from Definition 4.6.1 that $\Theta^{\vdash}$ is a theory. Suppose that Σ is a theory and $\Theta \subseteq \Sigma$. If $\alpha_1, \ldots, \alpha_n$ is a proof from Θ, arguing as in the proof of Lemma 4.5.1, we see that $\alpha_n \in \Sigma$. Therefore, $\Theta^{\vdash} \subseteq \Sigma$. The proof of (ii) trivially follows from (i). To prove (iii), note that if $\beta \in |\alpha|$ then $\alpha \to \beta \in \Theta^{\vdash}$ and, by (T2), $\beta \in \Theta$. □

Let $\{\Theta_i\}_{i \in I}$ be a nonempty family of theories. It is easy to see that $\bigcap_{i \in I} \Theta_i$ is a theory; further, from the above proposition we immediately get that $(\bigcup_{i \in I} \Theta_i)^{\vdash}$ is the smallest theory containing Θ_i for all $i \in I$. Thus, the set of theories of Łukasiewicz infinite-valued calculus, ordered by inclusion, is a complete lattice. We shall denote this lattice by **Theo**.

Comparing Definitions 4.2.6 and 4.6.1 we immediately get

Theorem 4.6.3 *The correspondence*

(4.28) $\Theta \mapsto |\Theta| =_{def} \{|\alpha| \in \mathcal{L} \mid \alpha \in \Theta\}$

defines an isomorphism from the lattice **Theo** *onto the lattice of implicative filters of the Lindenbaum algebra* $\mathcal{L}$. *The inverse isomorphism is given by*

(4.29) $F \mapsto \{\alpha \in \mathbf{Form} \mid |\alpha| \in F\}$. □

An important result of classical propositional calculus is the *Deduction Theorem.* The next proposition gives a version of this theorem for the Łukasiewicz infinite-valued propositional calculus.

Proposition 4.6.4 *For arbitrary formulas* α *and* β, *and for each* $\Theta \subseteq$ **Form** *we have*

$$\beta \in (\Theta \cup \{\alpha\})^{\vdash}$$

if, and only if, there is an integer $n \geq 1$ *such that* $(\alpha^n \to \beta) \in \Theta^{\vdash}$.

Proof: Since $\beta \in (\Theta \cup \{\alpha\})^{\vdash}$ iff $|\beta| \in |(\Theta \cup \{\alpha\})^{\vdash}|$, the result follows from Proposition 4.2.9 and 4.6.3. □

Corollary 4.6.5 $\beta \in \{\alpha\}^{\vdash}$ *iff there is an integer* $n \geq 1$ *such that* $\alpha^n \to \beta$ *is a tautology.*

Proof: By Proposition 4.2.9(ii). □

The next theorem gives the desired necessary and sufficient condition for $\Theta^{\models}$ to coincide with $\Theta^{\vdash}$.

Theorem 4.6.6 *For each* $\Theta \subseteq$ **Form**, $\Theta^{\models} = \Theta^{\vdash}$ *iff* $|\Theta^{\vdash}|$ *is an intersection of maximal implicative filters of the Lindenbaum algebra* $\mathcal{L}$.

Proof: For each valuation $\nu\colon \mathbf{Form} \to [0,1]$ we have

(4.30) $|\nu^{-1}(\{1\})| = h_\nu^{-1}(\{1\})$,

where $h_\nu\colon \mathcal{L} \to [0,1]$ is the homomorphism defined in Lemma 4.5.6. We next observe that whenever α does not belong to $\Theta^{\models}$ there exists a valuation $\nu_\alpha\colon \mathbf{Form} \to [0,1]$ such that $\nu_\alpha(\beta) = 1$ for all $\beta \in \Theta$ and $\nu_\alpha(\alpha) < 1$. Therefore,

$$\Theta^{\models} = \bigcap_{\alpha \notin \Theta^{\models}} \nu_\alpha^{-1}(\{1\}),$$

whence, by Theorem 4.6.3 and equation (4.30),

$$|\Theta^{\models}| = \bigcap_{\alpha \notin \Theta^{\models}} |\nu_\alpha^{-1}(\{1\})| = \bigcap_{\alpha \notin \Theta^{\models}} h_{\nu_\alpha}^{-1}(\{1\}).$$

Consequently, if $\Theta^{\models} = \Theta^{\vdash}$, then $|\Theta^{\vdash}|$ is an intersection of maximal implicative filters of $\mathcal{L}$.

Conversely, suppose there is a family $\{M_i\}_{i\in I}$ of maximal implicative filters of $\mathcal{L}$ such that $|\Theta^{\vdash}| = \bigcap_{i\in I} M_i$. Since for each $i \in I$ the quotient MV-algebra $\mathcal{L}/M_i$ is simple, there is a homomorphism $h_i: \mathcal{L} \to [0,1]$ such that $M_i = h_i^{-1}(\{1\})$. By Lemma 4.5.6, for each $i \in I$ there is a valuation ν_i: **Form** $\to [0,1]$ such that $h_i = h_{\nu_i}$. Hence, again using Theorem 4.6.3 and (4.30), we conclude that

$$\Theta^{\vdash} = \bigcap_{i\in I} \nu_i^{-1}(\{1\}).$$

Thus, whenever $\alpha \notin \Theta^{\vdash}$ there is $j \in I$ such that $\nu_j(\alpha) < 1$. It follows that $\alpha \notin \Theta^{\models}$. In conclusion, $\Theta^{\models} \subseteq \Theta^{\vdash}$, and by Lemma 4.5.1, $\Theta^{\vdash} = \Theta^{\models}$, as required. □

Every implicative filter F is an intersection of maximal implicative filters if, and only if, $\neg F$ is an intersection of maximal ideals. Since by Theorem 4.4.1, $\mathcal{L} \cong Free_\omega$, the example given in Chapter 3 of an ideal of $Free_\omega$ which is not an intersection of maximal ideals, immediately yields an example of a set $\Theta \subseteq$ **Form** such that $\Theta^{\models} \not\subseteq \Theta^{\vdash}$.

On the other hand, if Θ is a nonempty *finite* set of formulas, say $\Theta = \{\alpha_1, \ldots, \alpha_k\}$, from (4.8) together with Theorem 4.6.3, it follows that

$$\Theta^{\vdash} = \{\alpha_1 \odot \cdots \odot \alpha_k\}^{\vdash}.$$

Then $|\Theta^{\vdash}|$ is a principal (i.e., a one-generated) implicative filter of $\mathcal{L}$. An application of Theorems 3.4.9 and 4.6.6 yields

Theorem 4.6.7 *For each finite set* $\Theta \subseteq$ **Form**, $\Theta^{\vdash} = \Theta^{\models}$. □

Definition 4.6.8 For each theory Θ, the MV-algebra $\mathcal{L}/|\Theta|$ is called *the Lindenbaum algebra of* Θ, and is denoted by $\mathcal{L}(\Theta)$.

Note that $\mathcal{L} \cong \mathcal{L}/|\emptyset^{\vdash}|$. Therefore, $\mathcal{L}$ is the Lindenbaum algebra of the theory $\emptyset^{\vdash}$. From Theorem 4.6.6, together with our criterion for semisimplicity, (Theorem 3.4.9), for any set of formulas Θ we also have: $\Theta^{\vdash} = \Theta^{\models}$ iff $\mathcal{L}(\Theta)$ is semisimple.

Theorem 4.6.9 *Up to isomorphism, every countable MV-algebra A is the Lindenbaum algebra of some theory in the infinite-valued calculus of Łukasiewicz built from the variables* $X_0, X_1, \ldots$.

Proof: Let $g: \{|X_0|, |X_1|, \ldots\} \to A$ be an arbitrary map from the set of logical equivalence classes of propositional variables, onto A. By Proposition 4.5.5, g can be extended to a (unique) surjective homomorphism $h: \mathcal{L} \to A$. Let $F = \{|\alpha| \in \mathcal{L} \mid h(|\alpha|) = 1\}$. Then by Theorem 4.6.3, F corresponds to a theory Θ, and from $A \cong \mathcal{L}/F$ we conclude $A = \mathcal{L}(\Theta)$. $\square$

Remark: A straightforward generalization of the above theorem shows that every MV-algebra arises as the Lindenbaum algebra of some theory, provided sufficiently many variables are available in the alphabet.

The logical counterpart of the *word problem for finitely presented MV-algebras*, mentioned in the Introduction, has the following uniformly positive solution, yielding a generalization of Corollary 4.5.3:

Theorem 4.6.10 *There is a Turing machine $\mathcal{U}$ such that, over any input (Θ, ψ), where Θ is a finite set of formulas and ψ is a formula, $\mathcal{U}$ decides in a finite number of steps whether or not ψ is a syntactic (= semantic) consequence of Θ.*

Proof: By (4.8) and Theorem 4.6.7 we can safely assume that Θ only contains a single formula $\theta = \theta(X_1, \ldots, X_n)$. In light of Proposition 4.2.9(ii), we can effectively enumerate all pairs (θ, ψ) such that ψ is a consequence of θ.

Conversely, we shall now describe an effective enumeration of all pairs (θ, ψ) such that formula ψ is *not* a consequence of θ. Arguing as in the proof of Proposition 4.1.5, first of all we shall effectively enumerate all triplets $(\mathbf{x}, \theta, \varphi)$ such that

(i) $\mathbf{x}$ is an m-tuple of rational numbers in $[0, 1]$ $(m \geq n)$, and,

(ii) letting f_θ and f_φ be the McNaughton functions corresponding to θ and φ via Propositions 3.1.8 and 4.5.5, we have $f_\theta(\mathbf{x}) = 1$ and $f_\varphi(\mathbf{x}) < 1$.

By Theorem 4.6.7, the existence of such rational point $\mathbf{x}$ is a sufficient condition for φ not to be a consequence of θ.

Claim. The existence of $\mathbf{x} \in [0,1]^m \cap \mathbf{Q}$ is also a necessary condition for φ not to be a consequence of θ.

As a matter or fact, assume φ is not a consequence of θ, and let $\mathbf{y} \in [0,1]^m$ be an m-tuple of *real* numbers such that $f_\theta(\mathbf{y}) = 1$ and $f_\varphi(\mathbf{y}) < 1$. It is sufficient to argue in case $m = n$. Let $\hat{f}_\varphi$ be the restriction of f_φ to the set

$$Z =_{def} \{\mathbf{z} \in [0,1]^n \mid f_\theta(\mathbf{z}) = 1\}.$$

By Proposition 3.1.8 f_φ is a continuous, piecewise linear function, each piece having integer coefficients. Arguing as in Proposition 3.3.1, we see that the minimum value of $\hat{f}_\varphi$ is attained at some point $\mathbf{z} \in Z$ with rational coordinates. Our claim is settled.

To complete the proof, from the above list of triplets $(\mathbf{x}, \theta, \varphi)$ it is now easy to construct the required effective enumeration of all pairs (θ, φ) such that φ is not a consequence of θ. A routine argument now yields the desired Turing machine $\mathcal{U}$ deciding whether φ is a consequence of θ. □

4.7 Bibliographical remarks

The paper [139] is a basic reference for Łukasiewicz infinite-valued calculus. The equivalence between MV-algebras and Wajsberg algebras was first proved by Rodríguez in [212] (see also Font, Rodríguez and Torrens [86], and Komori [130]).

The crucial distinction between syntactic and semantic consequence relations in the infinite-valued calculus is due to Wójcicki [243] and [244, Theorem 4.3.4]. The fact that the two notions coincide for finitely axiomatizable theories was first proved by Hay in [113]. A related result by Rose is in [215, Lemma 1]. A geometric visualization of the two notions, using the differential structure of McNaughton functions was given in [170].

For the proof of the completeness theorem, following Chang [38] we use examples of proofs already present in [216]. Historically, the original list of axioms conjectured by Łukasiewicz to be complete for the infinite-valued calculus also included the axiom

$$((\alpha \to \beta) \to (\beta \to \alpha)) \to (\beta \to \alpha),$$

whose redundancy was proved simultaneously by Meredith [154] and Chang [37].

Chapter 5

Ulam's game

The crucial problem of interpreting n truth values when $n > 2$ was investigated, among others, by Łukasiewicz himself. As shown in this chapter, a simple interpretation is given by *Ulam game*, the variant of the game of Twenty Questions where $n - 2$ lies, or errors, are allowed in the answers. The case $n = 2$ corresponds to the traditional game without lies. The game is originally described by Ulam on page 281 of his book [235] as follows:

> Someone thinks of a number between one and one million (which is just less than 2^{20}). Another person is allowed to ask up to twenty questions, to each of which the first person is supposed to answer only yes or no. Obviously the number can be guessed by asking first: Is the number in the first half million? then again reduce the reservoir of numbers in the next question by one-half, and so on. Finally the number is obtained in less than $\log_2(1000000)$. Now suppose one were allowed to lie once or twice, then how many questions would one need to get the right answer?

5.1 Questions and answers

Let us inspect a round of Ulam game: initially, the two players agree to fix a nonempty finite set S of numbers, called the *search space*, and an integer $m \geq 0$. Then the first player chooses a number $x \in S$, and the second player must find the unknown x, by asking the smallest possible number of questions, to each of which the first player can only answer "yes" or "no", being allowed at most m errors/lies in his answers. By definition, a *question* is a subset of S: thus for instance, the question

is x an even number ?

is identified with the set of all even numbers in S.

We can safely assume Pinocchio to be the first player, and identify ourselves with the second player. Pinocchio's *answers* are propositions of either form "yes, (x is even)", or "no, (x is odd)". Our state of knowledge about x is uniquely determined by the conjunction of these answers. The latter, in general, do not obey the rules of classical logic, for

(a) *The conjunction of two equal answers to the same repeated question need not be equivalent to a single answer.* Thus, the classical idempotence principle fails. To see this, let us assume that Pinocchio can lie at most once. Suppose we ask twice the following question "is x even?". If the answer is "yes" in both cases, then x must be even. However, after the first answer we are not certain that x is even.

(b) *The conjunction of two opposite answers to the same repeated question need not lead to contradiction.* To see this, again using the above example, if Pinocchio answers "yes" to the first question, and "no" to the second, then we cannot conclude that S is empty.

In the traditional error-free game, our knowledge on x is usually represented by a function $\# : S \to \{0,1\}$ giving, for every $z \in S$, the number $a = \#z$ of falsified answers, where all $a \geq 1$ are collapsed to 1. Stated otherwise, for each $z \in S$, $\#z = 0$ iff z does not falsify any answer; $\#z = 1$ iff z falsifies at least one answer.

In Ulam game with m lies, our knowledge is also given by a function $\sigma : S \to \{0, 1, \ldots, m, m+1\}$, counting the number $a = \sigma(z)$ of answers falsified by z, where all $a \geq m+1$ are collapsed to $m+1$: thus $\sigma(z) = 0, 1, \ldots, m$ or $m+1$, according as z falsifies $0, 1, \ldots, m$ or $\geq m+1$ answers, respectively.

5.2 Dynamics of states of knowledge

Let us assume that, after receiving a certain number of answers from Pinocchio, our knowledge is represented by the function $\sigma : S \to$

$\{0,\ldots,m+1\}$. Let D be an arbitrary subset of S, and $\overline{D} = S\backslash D$ its complementary subset. Suppose our question "does x belong to D?", now receives from Pinocchio the answer "yes". Then how should we represent our new knowledge σ' ? Let the $(m+2)$-tuple $(S_0,\ldots,S_{m+1})$, be defined by

$$S_i =_{def} \{z \in S \mid \sigma(z) = i\},$$

($i = 0,\ldots,m+1$). There is an obvious correspondence between σ and $(S_0,\ldots,S_{m+1})$, generalizing the correspondence between subsets and characteristic functions. The knowledge $\sigma' = (S'_0,\ldots,S'_{m+1})$, obtained from σ as a consequence of the positive answer to question D is given by

$$(5.1) \quad S'_0 = S_0 \cap D, \qquad S'_{m+1} = S_{m+1} \cup (S_m \cap \overline{D})$$

and, for each $i = 1,\ldots,m$,

$$(5.2) \quad S'_i = (S_i \cap D) \cup (S_{i-1} \cap \overline{D}).$$

As a matter of fact, whenever $y \in S$ is a valid candidate for the unknown x, the number $\sigma(y)$ of answers falsified by y will remain constant iff y satisfies D iff $y \in D$; on the other hand, $\sigma(y)$ must be increased by one iff y falsifies D iff $y \notin D$. One can similarly deal with the negative answer to question D; such answer has precisely the same effect as a positive answer to the opposite question $\overline{D}$.

Formulas (5.1) and (5.2) describing the dynamics of Ulam game acquire a particularly simple form if, instead of assigning to each $y \in S$ the number $\sigma(y)$ of falsified answers, we assign the *truth value* $\tau(y)$ given by

$$(5.3) \quad \tau(y) =_{def} 1 - \sigma(y)/(m+1).$$

Intuitively, *the truth value $\tau(y)$ measures, in units of $m+1$, how distant y is from the condition of falsifying too many answers.* More precisely,

$$(5.4)\ \tau(y) = \begin{cases} 0, & \text{if } y \text{ falsifies } \geq\ m+1 \text{ answers} \\ 1/(m+1), & \text{if } y \text{ falsifies } m \text{ answers} \\ \cdots & \cdots \quad \cdots \quad \cdots \\ m/(m+1), & \text{if } y \text{ falsifies one answer} \\ 1, & \text{if } y \text{ falsifies no answer.} \end{cases}$$

Any function

$$\tau : S \to \{0, \frac{1}{m+1}, \ldots, \frac{m}{m+1}, 1\}$$

is called a *state of knowledge* in Ulam game over S with m lies/errors. In accordance with this notation, the *initial* state is the constant function 1 over S. At the other extreme, the constant function 0 is the *incompatible state*, in which every element of S falsifies $m+1$ answers, or more.

For every question $D \subseteq S$, the *positive* answer

$$D^{yes} : S \to \{m/(m+1), 1\}$$

is naturally defined by the following stipulation:

$$(5.5)\ D^{yes}(z) = \begin{cases} 1, & \text{if } z \in D \\ 1 - 1/(m+1), & \text{if } z \in \overline{D}. \end{cases}$$

We similarly define the *negative* answer D^{no} by the stipulation $D^{no} = \overline{D}^{yes}$. In other words, for each $z \in S$:

$$(5.6)\ D^{no}(z) = \begin{cases} 1, & \text{if } z \notin D \\ 1 - 1/(m+1), & \text{if } z \in D. \end{cases}$$

The proof of the following proposition is an immediate consequence of (5.3)-(5.6), recalling from (1.2) the definition of Łukasiewicz conjunction:

Proposition 5.2.1 *Let $t \geq 1$ be an integer. For every $i = 1, \ldots, t$, let $D_i \subseteq S$ and $b(i) \in \{yes, no\}$. Let τ be the state of knowledge arising from the sequence of answers $D_1^{b(1)}, D_2^{b(2)}, \ldots, D_t^{b(t)}$. Then*

$$\tau = D_1^{b(1)} \odot \ldots \odot D_t^{b(t)},$$

where $\odot$ is pointwise Łukasiewicz conjunction in the MV-algebra $Ł_{m+2}$. □

5.3 Operations on states of knowledge

Once equipped with pointwise Łukasiewicz conjunction, the set $K_{S,m}$ of states of knowledge in Ulam game over the search space S with m lies/errors becomes an abelian monoid with neutral element 1 (the initial state). While Pinocchio's answers may be false/erroneous, their conjunction is all we know about the unknown number x. As the number of answers increases, our state of knowledge becomes sharper and sharper. To account for the natural order between states of knowledge, let us agree to write

$$\tau' \leq \tau''$$

(read τ' is *sharper* than τ'', or τ'' is *coarser* than τ') iff $\tau'(y) \leq \tau''(y)$ for all $y \in S$.

For every state $\tau \in K_{S,m}$ there is a coarsest state $\neg\tau$ that is *incompatible* with τ, in the sense that $\tau \odot \neg\tau = 0$. Naturally, $\neg\tau = 1 - \tau$. We can regard $\neg$ as a form of negation, without any fear of confusion with (5.5) and (5.6). In particular, our analysis in Section 5.1 (b) does not apply to the $\neg$ operation.

Using the operations $\neg$ and $\odot$ we can express the natural order between states of knowledge, by writing $\tau \odot \neg\sigma = 0$ instead of $\tau \leq \sigma$.

Since the operations $0, \neg, \oplus$ and $1, \neg, \odot$ are mutually interdefinable, by a slight abuse of terminology, let us refer to $\mathcal{K}_{S,m} = \langle K_{S,m}, 1, \neg, \odot \rangle$ as the MV-algebra of *states of knowledge* in Ulam game over S with m lies.

We say that an equation $\phi(X_1, \ldots, X_n) = \psi(X_1, \ldots, X_n)$ is *absolute* iff it is valid whenever the variables X_i are replaced by arbitrary states of knowledge τ_i in any possible MV-algebra $\mathcal{K}_{S,m}$, arising from a Ulam game with finite search space S and any number m of lies.

The associativity and commutativity of $\odot$, as well as the involutive property of $\neg$ are examples of absolute equations. By contrast, the idempotence equation $X \odot X = X$, precisely holds in case $m = 0$. The following result characterizes absolute equations:

Theorem 5.3.1 *Let the terms ϕ and ψ be obtained from the variables $X_1, \ldots, X_n$ and the constant 1 by a finite number of applications of the operations $\neg$ and $\odot$. Then the following conditions are equivalent for the equation $\phi = \psi$:*

(i) The equation is absolute;

(ii) The equation holds in every algebra $\mathcal{K}_{S,m}$ with S a one-element set, and m ranging over all integers ≥ 0;

(iii) The equation holds when its variables range over all rational numbers in the unit interval $[0,1]$, with the natural MV-operations;

(iv) The equation holds in every MV-algebra.

Proof: The equivalence (i) $\Leftrightarrow$ (ii) follows from Lemma 1.4.5(iii), since $\mathcal{K}_{S,m}$ is finite product of chains L_{m+2}. For the equivalence (ii) $\Leftrightarrow$ (iii) it is sufficient to note that if an equation fails in $\mathbf{Q} \cap [0,1]$, then it fails in some finite subalgebra L_{m+2}, as can be seen by inspection of the proof of Chang's completeness theorem 2.5.3. Finally, the equivalence (iii) $\Leftrightarrow$ (iv) is a reformulation of Chang's completeness theorem. □

From 4.5.3 we now get

Corollary 5.3.2 *There is an effective procedure (for definiteness, a Turing machine) to decide whether an equation is absolute.* □

5.4 Bibliographical remarks

If all questions are asked at the outset, independently of the answers, then finding an optimal strategy in Ulam game with m lies amounts to finding an optimal m-error correcting code—a very difficult task indeed, already for small $m \geq 2$ (see [140]). However, Ulam game allows one to *adapt* the tth question using the information received from the first $t-1$ answers. Thus, Ulam game is a chapter of Berlekamp's theory of communication with feedback [23]. As a matter of fact, when one studies searching strategies in Ulam game it is immaterial whether wrong answers are due to lies, errors or distortion. Optimal strategies for Ulam game are described in several papers (see, e.g., [202], [62], [107], [174], [225], [194], [6], [43]).

In [188] one can find a discussion of the close relationship between Ulam game and the Maximum Satisfiability problem—the problem whether a set of n boolean clauses contains a satisfiable subset of cardinality $n-m$.

For the logical interpretation of Ulam game we have been closely following [175] and [179]. The various representation theorems scattered throughout this book allow one to regard MV-algebras as algebras of states of knowledge in generalized Ulam games—where the number of lies may depend on the secret number $x \in S$, S itself may be infinite, truth values may be irrational numbers in $[0,1]$, or even nonstandard real numbers.

Chapter 6

Lattice-theoretical properties

In this chapter we study properties that are strongly related to the lattice structure of MV-algebras. We start by considering relations between the ideals of an MV-algebra A and the ideals of the lattice $\mathbf{L}(A)$. A stonean ideal of a bounded distributive lattice L is an ideal generated by complemented elements of L. We shall show that the minimal prime lattice ideals of $\mathbf{L}(A)$, as well as the stonean ideals of $\mathbf{L}(A)$, are always ideals of A.

An MV-algebra A is called hyperarchimedean iff all its ideals are stonean ideals of $\mathbf{L}(A)$. Hyperarchimedean MV-algebras have many features in common with boolean algebras. For instance, an MV-algebra is hyperarchimedean iff all its prime ideals are maximal. Further, every hyperarchimedean MV-algebra A is semisimple, and so is every homomorphic image of A. We shall describe the most general hyperarchimedean MV-algebra with one generator.

We shall also investigate the role of the complemented elements of $\mathbf{L}(A)$ in the direct product decompositions of A, and we shall introduce boolean products of MV-algebras as generalizations of finite direct product decompositions.

An MV-algebra is said to be complete iff its underlying lattice is closed under infinite sups and infs. The structure of complete MV-algebras explicitly depends on the underlying lattice structure, and every complete MV-algebra A is semisimple. We shall also prove that every complete MV-algebra A has a decomposition $A = A_1 \times A_2 \times A_3$, where A_1 is complete and atomic, A_2 is a power of $[0, 1]$, and A_3 is complete and without linear factors.

6.1 Minimal prime ideals

As usual, by an *ideal of a lattice* L with 0 and 1 we mean a subset I of L satisfying the following conditions:

(IL1) $0 \in I$;

(IL2) If $x \in I$ and $y \leq x$ then $y \in I$;

(IL3) If $x, y \in I$ then $x \vee y \in I$.

For every element $z \in L$, the *principal ideal* $(\, z\,]$ is defined by

(6.1) $(\, z\,] = \{x \in L \mid x \leq z\}$.

We say that I is *proper* iff $I \neq L$. We say that I is *prime* iff I is proper and for any two $x, y \in L$, if $x \wedge y \in I$ then either $x \in I$ or $y \in I$.

Every ideal of an MV-algebra A is also an ideal of the underlying lattice $\mathbf{L}(A)$. To see that the converse does not hold in general, let z be a nonboolean element of A. Then $(\, z\,]$ is an ideal of the lattice $\mathbf{L}(A)$, but is not an ideal of the MV-algebra A. As a matter of fact, by Corollary 1.5.6, an MV-algebra A is a boolean algebra iff every ideal of the lattice $\mathbf{L}(A)$ is an ideal of A.

Lemma 6.1.1 *Let J be an ideal of an MV-algebra A. Then J is a prime ideal of A iff J is a prime ideal of the underlying lattice* $\mathbf{L}(A)$.

Proof: Suppose that J is a prime ideal of the MV-algebra A and let $x, y \in A$ be such that $x \wedge y \in J$. By hypothesis we can safely assume $x \ominus y \in J$. Hence $(x \wedge y) \oplus (x \ominus y) \in J$. Moreover, from $(x \wedge y) \oplus (x \ominus y) = (x \oplus (x \ominus y)) \wedge (y \oplus (x \ominus y)) = (x \oplus (x \ominus y)) \wedge (x \vee y) \geq x \wedge (x \vee y) = x$, we get $x \in J$. Thus, J is a prime ideal of $\mathbf{L}(A)$.

Conversely, assume J to be a prime ideal of $\mathbf{L}(A)$, and let $x, y \in A$. Since by Proposition 1.1.7, $(x \ominus y) \wedge (y \ominus x) = 0 \in J$, then we must either have $x \ominus y \in J$ or $y \ominus x \in J$, and, since by assumption J is an ideal of A, then J is a prime ideal in A. □

Definition 6.1.2 An ideal K of an MV-algebra A (resp., of $\mathbf{L}(A)$) is said to be *minimal prime* iff it is a prime ideal of A (resp., of $\mathbf{L}(A)$), and whenever $I \subseteq K$ is a prime ideal of A, (resp., of $\mathbf{L}(A)$), then $I = K$.

Theorem 6.1.3 *Let A be an MV-algebra, and J be a proper ideal of the underlying lattice $\mathbf{L}(A)$. For each $z \in A$ let*

$$J_z = \{x \in A \mid z \ominus x \notin J\} \quad \text{and} \quad K = K(J) = \bigcap\{J_z \mid z \in A \backslash J\}.$$

Then $K(J)$ is an ideal of A and $K(J) \subseteq J$. If, in addition, J is a prime ideal of $\mathbf{L}(A)$, then $K(J)$ is a prime ideal of A.

Proof: Since $1 \in A \backslash J$, K is well defined. We shall prove that K is an ideal of the MV-algebra A. Trivially, $0 \in K$. If $t \leq u \in K$, then $t \in K$. Indeed, for each $z \in A \backslash J$ from $z \ominus u \notin J$ we get a fortiori, $z \ominus t \notin J$ (because, $z \ominus t \geq z \ominus u$). Thus $t \in K$, as required. Assume now $x, y \in K$, with the intent of proving $x \oplus y \in K$. Pick an element $z \in A \backslash J$. Then from $x \in K$ we get $z \ominus x \in A \backslash J$, and hence $y \in K$ implies $(z \ominus x) \ominus y \notin J$. Thus,

$$z \ominus (x \oplus y) = z \odot (\neg x \odot \neg y) = (z \ominus x) \ominus y \notin J,$$

whence $x \oplus y \in K$. We have thus shown that K is an ideal of A.

In order to prove that $K \subseteq J$, it is enough to observe that $z \ominus z = 0 \in J$; thus, whenever $z \notin J$, we must have $z \notin K$.

Claim: For each $z \in A$, if $x \wedge y \in J_z$, then either $x \in J_z$ or $y \in J_z$.

As a matter of fact, we have the identities

$$(z \odot \neg x) \vee (z \odot \neg y) = z \odot (\neg x \vee \neg y) = z \ominus (x \wedge y) \notin J,$$

and since J is an ideal of $\mathbf{L}(A)$, we must either have $z \ominus x \notin J$ or $z \ominus y \notin J$, as required to settle our claim.

To conclude the proof, assume J to be a prime ideal of $\mathbf{L}(A)$. Assume $x \notin K$ and $y \notin K$. Then there exist elements $s, t \in A \backslash J$ such that $x \notin J_s$ and $y \notin J_t$. By our assumption about J, $s \wedge t \notin J$. If $x \wedge y \in K$ (absurdum hypothesis) then $x \wedge y \in J_{s \wedge t}$ and hence, by our claim, we must either have $x \in J_{s \wedge t}$ or $y \in J_{s \wedge t}$. From the inclusion $J_{s \wedge t} \subseteq J_s \cap J_t$ we obtain a contradiction in either case, thus showing that $x \wedge y \notin K$. We conclude that $K \subseteq J \neq A$ is a prime ideal of A. □

Corollary 6.1.4 *Let A be an MV-algebra and $\mathbf{L}(A)$ its underlying lattice. Then the set of minimal prime ideals of A coincides with the set of minimal prime ideals of $\mathbf{L}(A)$.*

Proof: This is an immediate consequence of Theorem 6.1.3, together with Lemma 6.1.1. □

Let us recall that a *dual ideal*, or *filter*, of a lattice L is a subset F of L satisfying the following conditions:

(F1) $1 \in F$;

(F2) If $x \in F$ and $x \leq y$, then $y \in F$

(F3) If $x \in F$ and $y \in F$, then $x \wedge y \in F$.

Theorem 6.1.5 *Let J be a minimal prime ideal of an MV-algebra A. Then for every $s \in J$ there exists $t \in A \backslash J$ such that $t \wedge s = 0$.*

Proof: By Corollary 6.1.4, J is a minimal prime ideal of $\mathbf{L}(A)$. Let $F = \{y \in A \mid \text{ there is } x \in A\backslash J \text{ such that } s \wedge x \leq y\}$. Then F is a filter of $\mathbf{L}(A)$ and $s \in F$. Moreover, for every $x \in A\backslash J$ we have $x \in F$, in symbols,

(6.2) $A\backslash J \subseteq F$.

Claim: If $0 \notin F$ then there exists a prime ideal P of $\mathbf{L}(A)$ such that $P \cap F = \emptyset$.

As a matter of fact, by our standing assumption, the family of all ideals of $\mathbf{L}(A)$ that are disjoint from F is nonempty. Applying Zorn's lemma, let P be maximal in this family. Then P will be a prime ideal of $\mathbf{L}(A)$. For otherwise, letting $a, b \notin P$ and $a \wedge b \in P$, by our assumption about P there are $p, q \in P$ such that $p \vee a \in F$ and $q \vee b \in F$. Then $(p \vee a) \wedge (q \vee b) \in F$. On the other hand, by distributivity,

$$(p \vee a) \wedge (q \vee b) = (p \wedge q) \vee (p \wedge b) \vee (a \wedge q) \vee (a \wedge b) \in P,$$

thus contradicting the disjointness of P and F. In conclusion, P is prime and the claim is settled.

By (6.2) we now get $P \subseteq J$ and $s \in J\backslash P$, thus contradicting the fact that J is a minimal prime ideal of $\mathbf{L}(A)$. It follows that $0 \in F$, whence there exists $t \in A\backslash J$ such that $s \wedge t = 0$, as required. □

6.2 Stonean ideals and archimedean elements

Definition 6.2.1 Let L be a lattice with 0 and 1. An ideal I of L is said to be *stonean* iff for every $x \in I$ there is a boolean element $z \in I \cap B(L)$ such that $x \leq z$.

Stated otherwise, an ideal I of L is stonean iff it is generated by the ideal $I \cap \mathbf{B}(L)$ in the boolean algebra $\mathbf{B}(L)$. For every $x \in L$, the ideal $(x]$ is stonean iff $x \in \mathbf{B}(L)$. The following is a generalization of Corollary 1.5.6.

Corollary 6.2.2 *For any MV-algebra A, every stonean ideal of the underlying lattice $\mathbf{L}(A)$ is automatically an ideal of A.*

Proof: Let J be a stonean ideal of $\mathbf{L}(A)$. We have only to show that J is closed under the $\oplus$ operation. Let $x, y \in J$. Then there are $s, t \in J \cap \mathbf{B}(A)$, such that $x \leq s$ and $y \leq t$. Since by Theorem 1.5.3(iv), $x \oplus y \leq s \oplus t = s \vee t \in J \cap \mathbf{B}(A)$, we obtain $x \oplus y \in J$. □

Definition 6.2.3 An element x of an MV-algebra A is said to be *archimedean* iff there is an integer $n \geq 1$ such that $x \oplus \cdots \oplus x$ (n times) is boolean.

Notation: Throughout this chapter, following the notation of (2.7) we shall write

$$n.x \quad \text{as an abbreviation of} \quad x \oplus \cdots \oplus x \ (n \text{ times}).$$

Corollary 6.2.4 *For every element x of an MV-algebra A, the following conditions are equivalent:*

(i) There is an integer $n \geq 1$ such that $\neg x \vee n.x = 1$;

(ii) There is an integer $n \geq 1$ such that $n.x = (n+1).x$;

(iii) x is archimedean;

(iv) $\langle x \rangle$ is a stonean ideal of $\mathbf{L}(A)$.

Proof: (i) $\Rightarrow$ (ii): If $1 = \neg x \vee n.x = \neg(x \oplus n.x) \oplus n.x$ then $n.x \leq (n+1).x = x \oplus n.x \leq n.x$.

(ii) $\Rightarrow$ (iii): If $n.x = (n+1).x$, then by induction we get $n.x = (n+k).x$, for all integers $k \geq 0$. In particular, $n.x = 2n.x$ and by Theorem 1.5.3(iv), $n.x \in \mathbf{B}(A)$.

(iii) $\Rightarrow$ (iv): Let $n_0 \geq 1$ be an integer such that $n_0.x \in \mathbf{B}(A)$. By Theorem 1.5.3(iv), $n_0.x = 2n_0.x = kn_0.x$ for all $k \geq 2$. Assume $y \in \langle x \rangle$. Then $y \leq n.x$ for some n, whence a fortiori, for suitably large k, we get $y \leq kn_0.x \in \mathbf{B}(A) \cap \langle x \rangle$. Thus $\langle x \rangle$ is a stonean ideal of L(A).

(iv) $\Rightarrow$ (i): By assumption, since $x \in \langle x \rangle$, there is $z \in \mathbf{B}(A) \cap \langle x \rangle$ such that $x \leq z$. Since $z \in \langle x \rangle$, there is an integer $n \geq 1$ such that $z \leq n.x$, whence, by Theorem 1.5.3(vi), $1 = \neg z \oplus n.x = \neg z \vee n.x \leq \neg x \vee n.x \leq 1$. □

6.3 Hyperarchimedean algebras

Definition 6.3.1 An MV-algebra A is said to be *hyperarchimedean* iff all its elements are archimedean.

Equivalently, by Corollary 6.2.4(ii), for every $x \in A$ there is an integer $n \geq 1$ such that $n.x = (n+1).x$; one then immediately sees that boolean algebras, finite MV-algebras and simple MV-algebras are examples of hyperarchimedean MV-algebras.

If the MV-algebra A is a chain, then $\mathbf{B}(A)$ is the two-element set $\{0,1\}$. In this case, a nonzero element x is archimedean iff $1 = n.x$ for some integer $n \geq 1$. Then by Theorem 3.5.1, an MV-chain is hyperarchimedean iff it is isomorphic to a subalgebra of $[0,1]$.

While the class of hyperarchimedean MV-algebras is closed under subalgebras, finite products and homomorphic images, this class is not closed under infinite products. As a matter of fact, let

$$(6.3) \qquad A = \prod_{n\geq 2} \mathbf{Ł}_n.$$

Then each $\mathbf{Ł}_n$ is hyperarchimedean, but the product MV-algebra A is not. To see this, let $f : \{2,3,4,\ldots\} \to \bigcup_{n\geq 2} \mathbf{Ł}_n$ be defined by $f(n) = 1/(n-1)$. From $0 < n.f(n+2) = nf(n+2) = n/(n+1) < 1$, we get

$n.f \notin \mathbf{B}(A)$, for each $n = 0, 1, 2, \ldots$, whence f is a nonarchimedean element of A, and A is not hyperarchimedean.

Theorem 6.3.2 *For any MV-algebra A the following conditions are equivalent:*

(i) A is hyperarchimedean;

(ii) Every ideal of A is a stonean ideal of $\mathbf{L}(A)$;

(iii) Every prime ideal of A is maximal;

(iv) Every prime ideal of A is minimal;

(v) Every ideal of A is an intersection of the maximal ideals containing it;

(vi) For every ideal J of A, A/J is semisimple.

Proof: (i) $\Rightarrow$ (ii): Let J be an ideal of A, and $x \in J$. By Corollary 6.2.4(iii) there is an integer $n > 0$ such that $n.x$ is a boolean element. Since $x \leq n.x \in J$, J is a stonean ideal of $\mathbf{L}(A)$.

(ii) $\Rightarrow$ (iii): Let P be a prime ideal in A and $P \subseteq J$ for some ideal J in A. Let $P_0 = P \cap \mathbf{B}(A)$, $J_0 = J \cap \mathbf{B}(A)$. Suppose $x \in J \backslash P$ (absurdum hypothesis). Since by hypothesis J is stonean, there is $z \in J_0$ such that $x \leq z$; it follows that $z \notin P_0$. Since P is prime, it is easy to see that P_0 is a maximal ideal in the boolean algebra $\mathbf{B}(A)$. Since J_0 strictly contains P_0, we must have $1 \in J_0$, whence $J \supseteq J_0$ is not a proper ideal, a contradiction showing that P is maximal.

(iii) $\Leftrightarrow$ (iv): Trivial.

(iv) $\Rightarrow$ (i): Let $0 \neq s \in A$, and $J^{\perp} = \{x \in A \mid x \wedge s = 0\}$. Then $J^{\perp}$ is an ideal of A. As a matter of fact, skipping all trivialities, whenever $x \wedge s = 0 = y \wedge s$, then $(x \oplus y) \wedge s = 0$; this is easily verified on all MV-chains, and hence by the subdirect representation theorem, it holds in general. Let K be the ideal generated by $J^{\perp}$ and s. By (1.9) we have

$$K = \{x \in A \mid x \leq n.s \oplus a, \text{ for some } n = 1, 2, \ldots \text{ and } a \in J^{\perp}\}.$$

Assume K to be a proper ideal (absurdum hypothesis). Then, by Corollary1.2.14 there is a prime ideal J in A containing K. Since by hypothesis, J is minimal, an application of Theorem 6.1.5 yields an element $t \in A\backslash J$ such that $s \wedge t = 0$. Since $J \supseteq K \supseteq J^{\perp}$, then $t \notin J^{\perp}$, a contradiction. We have shown that $1 \in K$. Thus, there is an integer $n \geq 1$ and an element $a \in J^{\perp}$ such that $n.s \oplus a = 1$. From $a \wedge s = 0$ and $\neg n.s \leq a$ it follows that $\neg n.s \wedge s = 0$, whence $1 = n.s \vee \neg s$. By Corollary 6.2.4(i), s is an archimedean element of A, and A is hyperarchimedean.

(iii) $\Leftrightarrow$ (v): This is a consequence of Corollary 1.2.14.

(v) $\Leftrightarrow$ (vi): By Lemma 3.6.6. □

By Corollary 3.6.8, since each hyperarchimedean MV-algebra A is semisimple, there is a compact Hausdorff space X such that A is isomorphic to a separating subalgebra of $Cont(X)$.

Proposition 6.3.3 *Let X be a compact Hausdorff space. A separating subalgebra A of $Cont(X)$ is hyperarchimedean iff $Z(f) = f^{-1}(\{0\})$ is an open subset of X for each $f \in A$.*

Proof: Suppose that A is hyperarchimedean, and for each $x \in X$, let O_x denote the ideal given by all functions in A that vanish over some open neighbourhood of x. The fact that A separates points implies that $V_{O_x} = \{x\}$, and then, from Theorem 3.4.3(iii) and Theorem 6.3.2(vi) it follows that $O_x = J_{V_{O_x}} = J_x$. Therefore, whenever $f \in A$ vanishes in some point $x \in X$, then f also vanishes over an open neighbourhood of x. This property clearly implies that $Z(f)$ is an open subset of X for each $f \in A$.

To prove the converse, we need the following

Claim: Let J be an ideal of A, and let $f \in A$. If f vanishes over some open set $U \supseteq V_J$, then $f \in J$.

As a matter of fact, since $V_J \cap (X \setminus U) = \emptyset$, for each $x \in X \setminus U$ there is a function $g_x \in J$ such that $g_x(x) > 0$. Since $X \setminus U$ is a compact subset of X, arguing as in the proof of Theorem 3.4.3(ii), we can find a function $g \in J$ such that $g(x) = 1$ for each $x \in X \setminus U$. Since $f(x) = 0 \leq g(x)$, for all $x \in U$, and $f(x) \leq 1 = g(x)$, for all $x \in X \setminus U$, then $f \leq g$. We conclude that $f \in J$, as required to settle our claim.

Suppose now that $Z(f)$ is open for each $f \in A$, and let J be a proper ideal of A. If $f \in J_{V_J}$, then $V_J \subseteq Z(f)$, and by the above claim, $f \in J$. Therefore $J = J_{V_J}$, and again by Theorem 3.4.3(iii) and Theorem 6.3.2(vi) we can conclude that A is hyperarchimedean. □

We shall now characterize hyperarchimedean MV-algebras with one generator. In the light of Theorem 3.6.7 and Proposition 3.1.5, we can safely focus attention on MV-algebras of restrictions of McNaughton functions, as follows:

Corollary 6.3.4 *Let $X \neq \emptyset$ be a closed subset of the real interval $[0,1]$, and let A be the subalgebra of $Cont(X)$ given by the restrictions to X of the functions in $Free_1$. Then A is hyperarchimedean iff every rational point of X is isolated.*

Proof: By Corollary 3.2.8, $Free_1$ coincides with the MV-algebra of all one-variable McNaughton functions. Suppose that A is hyperarchimedean and let r be a rational point in X. We can easily define a McNaughton function $f_r: [0,1] \to [0,1]$ such that $Z(f_r) = \{r\}$. Then from Proposition 6.3.3, we obtain that $\{r\} = Z(f_r) \cap X$ is open in X, i.e., r is isolated in X.

Conversely, suppose that all rational points in X are isolated, and let $f: [0,1] \to [0,1]$ be a McNaughton function. Clearly $Z(f|_X) = Z(f) \cap X$. Let $z \in Z(f) \cap X$. If z is rational, then $\{z\}$ is an open neighbourhood of z (in the topology of X) that is contained in $Z(f|_X)$. Suppose that z is irrational. Since f is a McNaughton function, there is an open interval U such that $z \in U \subseteq Z(f)$. Then $U \cap X$ is an open neighbourhood of z in X and is contained in $Z(f|_X)$. This shows that $Z(f|_X)$ is open in X, whence by Proposition 6.3.3, A is hyperarchimedean. □

Let X be a compact Hausdorff space. Following tradition, we say that a set $S \subseteq X$ is *clopen* iff S is simultaneously open and closed. $\emptyset$ and X are always clopen subsets of X. The clopen subsets of X equipped with union, intersection and complement, form a boolean algebra, denoted clop(X).

On the other hand, it is easy to see that

$$\mathbf{B}(Cont(X)) = \{f \in Cont(X) \mid \text{ for all } x \in X,\ f(x) \in \{0,1\}\}.$$

Therefore the correspondence $f \mapsto Z(f)$ defines an isomorphism from $\mathbf{B}(Cont(X))$ onto clop(X). More generally, let U be a clopen subset of

X and A be a separating subalgebra of $Cont(X)$. Since U is compact, for each $x \notin U$ there is $f_x \in A$ such that $f_x(x) = 1$ and $U \subseteq Z(f_x)$. Since $X \setminus U$ is also compact, we can find $f \in A$ such that $f(t) > \frac{1}{2}$ for each $t \in X \setminus U$ and $Z(f) = U$. Hence $g = 2.f \in \mathbf{B}(A)$ and $Z(g) = U$. In conclusion, we have proved that *the correspondence* $g \mapsto Z(g)$ *defines an isomorphism of* $\mathbf{B}(A)$ *onto* $\mathrm{clop}(X)$. In particular, for each separating subalgebra A of $Cont(X)$, we have $\mathbf{B}(A) \cong \mathbf{B}(Cont(X))$.

Let X be a compact Hausdorff space and A be a separating subalgebra of $Cont(X)$. If A is hyperarchimedean then, by Proposition 6.3.3, the complements of the zero sets of functions in A are clopen and, by the remark following Theorem 3.4.3, they form a basis of the topology of X. Compact Hausdorff spaces having a basis of clopen sets are called *boolean spaces* .

For every boolean algebra B let

$$\mathbf{X}(B)$$

be the set of maximal ideals of B, equipped with the topology having as an open basis the sets of the form $\sigma_B(a) = \{P \in \mathbf{X}(B) \mid a \notin P\}$, where a ranges over elements of B. Then $\mathbf{X}(B)$ is a boolean space. Furthermore, Stone's celebrated theorem states that $B \cong \mathrm{clop}(\mathbf{X}(B))$, and $\mathbf{X}(B)$ is uniquely determined, up to homeomorphism. Since $\mathbf{B}(Cont(X)) \cong \mathrm{clop}(X)$, we conclude that *a Hausdorff compact space* X *is a boolean space iff* $\mathbf{B}(Cont(X))$ *is a separating subalgebra of* $Cont(X)$. We have proved:

Corollary 6.3.5 *Let X be a compact Hausdorff space. Then the MV-algebra $Cont(X)$ has a hyperarchimedean separating subalgebra A iff X is a boolean space. In this case, X is homeomorphic to* $\mathbf{X}(\mathbf{B}(A))$. □

It is worthwhile to point out that the fact that X is a boolean space need not imply that $Cont(X)$ is a hyperarchimedean MV-algebra, as the following example shows.

Example: Let X be Cantor's ternary subset of $[0, 1]$. It is well known that X, with the topology inherited from $[0, 1]$ is a boolean space having no isolated points. Since X contains rational points, by Corollary 6.3.4, the separating subalgebra of $Cont(X)$ formed by the restriction of McNaughton functions to X is not hyperarchimedean, whence $Cont(X)$ is not hyperarchimedean.

6.4 Direct products

Let A be an MV-algebra. For each $z \in A$ let the functions $\neg_z : A \to A$ and $h_z : A \to A$ be defined by

(6.4) $h_z(x) = z \wedge x$ and $\neg_z x = z \wedge \neg x$.

Proposition 6.4.1 *For each nonzero element $b \in \mathbf{B}(A)$, $((b], \oplus, \neg_b, 0)$ is an MV-algebra and h_b is a homomorphism of A onto $(b]$ with $Ker(h_b) = (\neg b]$.*

Proof: By Corollary 1.5.6, $(b]$ is an ideal of A, and hence it is closed under the restriction of the operation $\oplus$ over $(b]$. We shall now show that $(b]$ is an MV-algebra. We first note that $(b]$ is closed under the operation $\neg_b$, and that for all $x \in (b]$, $\neg_b\neg_b x = x$. Further, by Theorem 1.5.3(vi), for every $x \in (b]$ we have $x \oplus \neg_b 0 = x \oplus b = x \vee b = b$. To establish the identity $\neg_b(x \oplus \neg_b y) \oplus x = \neg_b(\neg_b x \oplus y) \oplus y$, we first note that for any $y \in (b]$, $b \oplus y = b \vee y = b$. It follows that

$$\neg_b(\neg_b x \oplus y) \oplus y = \neg_b((b \wedge \neg x) \oplus y) \oplus y$$

$$= \neg_b((b \oplus y) \wedge (\neg x \oplus y)) \oplus y$$

$$= (b \wedge \neg(b \wedge (\neg x \oplus y))) \oplus y = ((b \wedge \neg b) \vee (b \wedge \neg(\neg x \oplus y))) \oplus y$$

$$= (b \oplus y) \wedge (\neg(\neg x \oplus y) \oplus y) = b \wedge (\neg(\neg x \oplus y) \oplus y).$$

By symmetry,

$$\neg_b(x \oplus \neg_b y) \oplus x = b \wedge (\neg(x \oplus \neg y) \oplus x) = b \wedge (\neg(\neg x \oplus y) \oplus y)$$

$$= \neg_b(\neg_b x \oplus y) \oplus y,$$

as required. The remaining verifications to establish that $(b]$ is an MV-algebra, are all trivial. In order to prove that h_b is a homomorphism, for all $x, y \in A$ we have $(x \wedge b) \oplus (y \wedge b) = ((x \wedge b) \oplus y) \wedge ((x \wedge b) \oplus b)$. Since, by Theorem 1.5.3(vi), $(x \wedge b) \oplus b = (x \wedge b) \vee b = b$, we get

$$(x \wedge b) \oplus (y \wedge b) = (x \oplus y) \wedge (b \oplus y) \wedge b = (x \oplus y) \wedge b.$$

We conclude that $h_b(x \oplus y) = h_b(x) \oplus h_b(y)$. The rest of the proof is trivial. □

Definition 6.4.2 For any partially ordered set X with minimum element 0, by an *atom* of X we mean an element $a \in X$ such that $a > 0$ and whenever $x \in X$ and $x \leq a$ then either $x = 0$ or $x = a$.

The following is an immediate consequence of the definition of $(b]$:

Proposition 6.4.3 *For every MV-algebra A and $0 < b \in \mathbf{B}(A)$ we have*

(i) The MV-algebras $(b]$ and $A/(\neg b]$ are isomorphic;

(ii) $(b]$ is a subalgebra of A iff $b = 1$ iff $(b] = A$;

(iii) $\mathbf{B}((b]) = (b] \cap \mathbf{B}(A)$. If in addition, $(b]$ is a chain, then b is an atom of the boolean algebra $\mathbf{B}(A)$. □

Remark: For each $0 < a \in A$, upon defining $x \oplus_a y = (x \oplus y) \wedge a$ and $\neg_a x = a \odot \neg x$, it follows that $((a], \oplus_a, \neg_a, 0)$ is an MV-algebra. However, if a is not a boolean element of A, then in general $(a]$ is not a homomorphic image of A. For instance, let $n \geq 3$ and $0 < k < n-1$. Then $k/(n-1)$ is an element of the Łukasiewicz chain $\mathbf{Ł}_n$ and $(k/(n-1)]$ is isomorphic to $\mathbf{Ł}_{k+1}$. Since $k+1 < n$ and $\mathbf{Ł}_n$ is simple, there is no homomorphism of $\mathbf{Ł}_n$ onto $(k/(n-1)]$. On the other hand, the existence of a homomorphism of A onto $(a]$ need not imply that a is a boolean element of A. As a matter of fact, for each $0 < a \in [0,1]$, the map $x \mapsto ax$ is an isomorphism of $[0,1]$ onto $(a]$. However, a is not a boolean element of $[0,1]$, unless a=1.

Notation: Given a nonempty family $\{a_i\}_{i \in I}$ of elements of an MV-algebra A, we write $\bigvee_{i \in I} a_i = 1$ iff 1 is the only upper bound of the family.

Lemma 6.4.4 *Let $\{A_i\}_{i\in I}$ be a nonempty family of MV-algebras and let $P = \prod_{i\in I} A_i$. Then there is a set $\{\delta_i \mid i \in I\} \subseteq \mathbf{B}(P)$ satisfying the following conditions:*

(i) $\bigvee_{i\in I} \delta_i = 1$;

(ii) $\delta_i \wedge \delta_j = 0$, *whenever* $i \neq j$;

(iii) each A_i *is isomorphic to* $(\,\delta_i\,]$.

Proof: For each $i \in I$, let $\delta_i: I \to \bigcup_{i\in I} A_i$ be defined by

$$\delta_i(j) = \begin{cases} 1 \in A_i & \text{if } j = i \\ 0 \in A_i & \text{if } j \neq i. \end{cases}$$

Then $\delta_i \in \mathbf{B}(\mathrm{P})$, $\delta_i \wedge \delta_j = 0$, for all $i \neq j$, and $\bigvee_{i\in I} \delta_i = 1$. Let $\pi_i: P \to A_i$ be the canonical projection. Then the kernel of $h_{\delta_i} : P \to (\,\delta_i\,]$ coincides with the kernel of π_i. As a matter of fact, by Corollary 6.4.1 we can write $Ker(h_{\delta_i}) = (\neg\delta_i\,] = \{f \in P \mid f(i) = 0\} = Ker(\pi_i)$. Thus, by Lemma 1.2.7, A_i is isomorphic to $(\,\delta_i\,]$. □

The above lemma has the following partial converse:

Lemma 6.4.5 *Let A be an MV-algebra. Let $b_1, \ldots, b_k$, $k \geq 2$ be elements in $\mathbf{B}(A)$ such that*

(i) $b_1 \vee \cdots \vee b_k = 1$, *and*

(ii) $b_i \wedge b_j = 0$ *for* $i \neq j$, $i, j = 1, \ldots, k$.

Then $A \cong (b_1] \times \cdots \times (b_k]$.

Proof: From (i) we immediately get $\bigcap_{i=1}^{k} (\neg b_i] = \{0\}$. Hence by Theorem 1.3.2 and Proposition 6.4.1, the function $h: A \to (b_1] \times \cdots \times (b_k]$ given by $h(a) = (a \wedge b_1, \ldots, a \wedge b_k)$, for each $a \in A$, is an embedding. Since by (ii), $h(a_1 \vee \cdots \vee a_k) = (a_1, \ldots, a_k)$, then h is also surjective, whence h is an isomorphism. □

Definition 6.4.6 An MV-algebra A is called *directly indecomposable* iff A is nontrivial, and whenever we can write $A \cong A_1 \times A_2$ then either A_1 or A_2 is trivial.

As an immediate consequence of Lemmas 6.4.4 and 6.4.5 we have:

Theorem 6.4.7 *An MV-algebra A is directly indecomposable if and only if* $\mathbf{B}(A) = \{0,1\}$. $\Box$

From Proposition 6.4.3(iii) and the above theorem we obtain that for each atom b of $\mathbf{B}(A)$, the MV-algebra $(b] \cong A/(\neg b]$ is directly indecomposable. This result can be generalized as follows: first of all, let us say that an ideal J of an MV-algebra A is a *maximal stonean ideal* iff J is a proper stonean ideal of A, and for each stonean ideal I of A, $I \neq J$ and $J \subseteq I$ imply $I = A$. In other words, $J \cap \mathbf{B}(A)$ is a prime ideal of $\mathbf{B}(A)$.

Corollary 6.4.8 *Let A be a nontrivial MV-algebra. For each maximal stonean ideal J of A, the quotient algebra A/J is directly indecomposable.*

Proof: Let $x \in A$ be such that $x/J \in \mathbf{B}(A/J)$. Then $x \wedge \neg x \in J$, and there is $z \in J \cap \mathbf{B}(A)$ such that $x \wedge \neg x \leq z$. Letting $y = x \vee z$, we get $y \wedge \neg y = (x \wedge \neg x \wedge \neg z) \vee (z \wedge \neg x \wedge \neg z) = 0$. It follows that $y \in \mathbf{B}(A)$, and since $J \cap \mathbf{B}(A)$ is a prime (= maximal) ideal of $\mathbf{B}(A)$, then either $y \in J$ or $\neg y \in J$. If $y \in J$, then $x \in J$, because $x \leq y$. Suppose $y \notin J$. Then $\neg y \in J$, and $\neg x \vee z = \neg y \vee z \in J$. Hence $\neg x \in J$. $\Box$

6.5 Boolean products of MV-algebras

As shown by Lemmas 6.4.4 and 6.4.5, boolean elements have an important role in direct product decompositions of MV-algebras. We will return to this topic in the next section, when considering complete MV-algebras. Now we are going to consider a special kind of subdirect product, where the index set is equipped with a boolean topology.

Definition 6.5.1 A *weak boolean product* of a family $\{A_x\}_{x \in X}$ of MV-algebras ($X \neq \emptyset$) is a subdirect product A of the given family, in such a way that X can be endowed with a boolean (i.e., totally disconnected, compact Hausdorff) topology having the following two properties:

(i) For all $f, g \in A$, the set $[\![f = g]\!] =_{def} \{x \in X \mid f(x) = g(x)\}$ is open in X;

(ii) Whenever Z is a clopen (closed-and-open) subset of X and $f, g \in A$ then $f|_Z \cup g|_{X \setminus Z} \in A$.

Replacing condition (i) by:

(i*) For all $f, g \in A$, the set $[\![f = g]\!]$ is clopen,

we obtain the notion of a *boolean product.*

Recalling the properties of the distance function, since $a = b$ is equivalent to $d(a, b) = 0$, the above conditions (i) and (i*) can be replaced, respectively, by:

(i') If $f \in A$, then $[\![f = 0]\!]$ is open in X

and

(i*') If $f \in A$, then $[\![f = 0]\!]$ is clopen in X.

In the following, when dealing with a weak boolean product A of a family $\{A_x\}_{x \in X}$, the dependence of A on the topology of X shall be tacitly understood. As shown by the following example, boolean products are a generalization of finite direct products.

Example: Let $\{A_x\}_{x \in X}$, for $X \neq \emptyset$ be a family of nontrivial MV-algebras, and let $A = \prod_{x \in X} A_x$. Then A is a weak boolean product of the family $\{A_x\}_{x \in X}$ if, and only if, X is finite. In this latter case, A is indeed a boolean product; further, the discrete topology on X is the only possible topology making A into a boolean product of the family $\{A_x\}_{x \in X}$. As a matter of fact, assuming X to be finite, then conditions (i*') and (ii) are trivially satisfied upon equipping X with the discrete topology. A moment's reflection shows that no other topology on X can make A into a boolean product of the $\{A_x\}_{x \in X}$. Conversely, assume X to be infinite, with the intent of proving that A cannot be made into a weak boolean product of the $\{A_x\}_{x\in X}$. For each $z \in X$ let the function $f_z: X \to \bigcup_{x\in X} A_x$ be defined by stipulating that, for each $x \in X$,

$$f_z(x) = \begin{cases} 0_z \in A_z & \text{if } x = z, \\ 1_x \in A_x & \text{if } x \neq z. \end{cases}$$

Since $f \in A$ and $[\![f_z = 0]\!] = \{z\}$, then, for condition (i) to hold it is necessary to endow X with the discrete topology. While this topology is Hausdorff and has a basis of clopen sets, it fails to make X into a compact space.

The following theorem should be compared with Lemmas 6.4.4 and 6.4.5, together with the above example. Recall the definition of $\mathbf{X}(C)$ for any boolean algebra C.

Theorem 6.5.2 *Let A be a weak boolean product of a family $\{A_x\}_{x\in X}$, $X \neq \emptyset$, of nontrivial MV-algebras. Let C be defined by*

(6.5) $\quad C = \{g \in A \mid g(x) \in \{0_x, 1_x\}$ *for each* $x \in X\}$.

Then C is a subalgebra of $\mathbf{B}(A)$ and we have:

(i) The correspondence $x \mapsto Q_x = \{g \in C \mid g(x) = 0\}$ is a homeomorphism from X onto $\mathbf{X}(C)$;

(ii) For each $x \in X$, A_x is isomorphic to $A/\langle Q_x\rangle$;

(iii) C coincides with $\mathbf{B}(A)$ iff all algebras A_x are directly indecomposable.

Conversely, if A is a nontrivial MV-algebra and C is a subalgebra of $\mathbf{B}(A)$, then A is isomorphic to a weak boolean product of the family $\{A/\langle Q\rangle\}_{Q\in\mathbf{X}(C)}$.

Proof: Suppose first that A is a boolean product of the family $\{A_x\}_{x\in X}$, and let C be as in (6.5). Then C is a subalgebra of $\mathbf{B}(A)$, and by Definition 6.5.1, $C \cong \mathbf{B}(X)$; arguing as in Theorem 3.4.3 one easily sees that the map $x \mapsto Q_x$ is a one-one correspondence of X onto $\mathbf{X}(C)$. As a matter of fact, this correspondence is continuous, because for each $x \in X$ and $g \in C$ we have $Q_x \in \sigma_C(g)$ iff $x \in X \setminus [\![g = 0]\!] = [\![\neg g = 0]\!]$, the latter being open by condition (i) in Definition 6.5.1. Hence, inverse images of basic open sets of $\mathbf{X}(C)$ are open in X. Now (i) follows from the well known fact that continuous bijections between compact Hausdorff spaces are homeomorphisms. Arguing as in the proof of Theorem 1.3.2, from the assumption that A is a subdirect product of the family $\{A_x\}_{x\in X}$, letting for each $x \in X$, $P_x = \{f \in A \mid f(x) = 0\} = Ker(\pi_x)$, it is easy to see that $A_x \cong A/P_x$. Hence to prove (ii) we need to show that $P_x = \langle Q_x \rangle$ for each $x \in X$. Suppose $f \in P_x$. Then $x \in [\![f = 0]\!]$, and since $x \mapsto P_x$ is a homeomorphism from X onto $\mathbf{X}(C)$, there is an element $g \in C$ such that $x \in [\![g = 1]\!] = [\![\neg g = 0]\!] \subseteq [\![f = 0]\!]$. Therefore, $\neg g \in P_x \cap C = Q_x$ and $f \leq \neg g$. Hence $P_x = \langle Q_x \rangle$, as required. If $C = \mathbf{B}(A)$, for each $x \in X$, $P_x = \langle Q_x \rangle$ is a maximal stonean ideal of A, and by Corollary 6.4.8, $A/\langle Q_x \rangle$ is directly indecomposable. On the other hand, if for some $x \in X$, A_x is not directly indecomposable then Q_x cannot be a prime ideal of $\mathbf{B}(A)$. Since Q_x is a prime ideal of C, the latter must be a proper subalgebra of $\mathbf{B}(A)$. Hence (iii) holds.

Conversely, assume that C is a subalgebra of $\mathbf{B}(A)$. Then, in the light of Theorem 1.3.2, in order to prove that A is a subdirect product of the family $\{\langle Q \rangle\}_{Q\in\mathbf{X}(C)}$, it suffices to prove the following:

Claim. The intersection of all the ideals of A that are generated by prime ideals of C coincides with the ideal $\{0\}$.

Let $0 \neq a \in A$. Applying Corollary 1.2.14 to the ideal $\{0\}$, we get a prime ideal P of A such that $a \notin P$. Since $P \cap C$ is a prime ideal of C, to complete the proof it suffices to show that $a \notin Q = \langle P \cap C \rangle$. Suppose that $a \in Q$ (absurdum hypothesis). Then there is $c \in P \cap C$ such that $a \leq c$. Moreover, $a \wedge \neg c = 0 \in P$, and since $a \notin P$, by Lemma 6.1.1 we obtain $\neg c \in P$. Hence, $1 = c \oplus \neg c \in P$, and $P = A$, a contradiction. This proves the claim.

To simplify the notation, we can safely identify A with its corresponding subalgebra of $\prod_{Q\in\mathbf{X}(C)} A/\langle Q \rangle$, and regard the elements of A as functions

$$f: X \to \bigcup_{Q\in\mathbf{X}(C)} A/\langle Q \rangle.$$

Suppose $f \in A$, and let $Q \in [\![f = 0]\!]$. Then $f \in \langle Q \rangle$, and there is $g \in Q$ such that $f \leq g$. It follows that

$$Q \in \sigma_C(\neg g) = [\![g = 0]\!] \subseteq [\![f = 0]\!],$$

whence condition (i) in Definition 6.5.1 holds. Finally, let Z be a clopen subset of X. Then there is $g \in C$ such that $Z = \sigma_C(g)$, and since for all f and h in A, $f|_Z \cup h|_{X \setminus Z} = (g \wedge f) \vee (\neg g \wedge h) \in A$, we conclude that (ii) in Definition 6.5.1 also holds. □

Corollary 6.5.3 *Each nontrivial MV-algebra is isomorphic to a weak boolean product of directly indecomposable MV-algebras.* □

Let A be an MV-algebra. Since A/J is an MV-chain iff J is a prime ideal, and MV-chains are directly indecomposable, we obtain:

Corollary 6.5.4 *A nontrivial MV-algebra A is a weak boolean product of MV-chains if and only if each maximal stonean ideal of A is a prime ideal of A.* □

Particular cases of MV-chains are given by simple MV-algebras, i.e., the subalgebras of $[0, 1]$. Since for any MV-algebra A, A/J is simple if and only if J is maximal ideal, we obtain:

Corollary 6.5.5 *An MV-algebra A is a weak boolean product of simple algebras iff each maximal stonean ideal of A is a maximal ideal of A.* □

Suppose that A is a weak boolean product of simple MV-algebras, and let $a \in A$ and $Q \in \mathbf{X}(\mathbf{B}(A))$ be such that $Q \in [\![a \neq 0]\!]$. Then a does not belong to the ideal $\langle Q \rangle$ generated by Q in A, and since by Corollary 6.5.5 $\langle Q \rangle$ is a maximal ideal of A, there are $c \in \langle Q \rangle$ and $n \in \mathbf{N}$ such that $c \oplus na = 1$, i.e., $\neg na \leq c$. Hence there is $b \in Q$ such that $\neg na \leq b$, whence $Q \in [\![b = 0]\!] \subseteq [\![a \neq 0]\!]$. Therefore $[\![a = 0]\!]$ is clopen for each $a \in A$. We have proved that *weak boolean products of simple MV-algebras are automatically boolean products.*

On the other hand, if P is a prime ideal of an MV-algebra A, then $\langle P \cap \mathbf{B}(A) \rangle$ is a maximal stonean ideal contained in P. Therefore, in the light of Theorem 6.3.2 we conclude that *each maximal stonean ideal of A is a maximal ideal of A iff A is hyperarchimedean.*

From the above remarks and Corollary 6.5.5, we get the following characterization of hyperarchimedean MV-algebras:

Corollary 6.5.6 *A nontrivial MV-algebra A is isomorphic to a boolean product of simple MV-algebras iff A is hyperarchimedean.* □

The example given at the end of Section 6.3 shows that in general, $Cont(X)$, for a boolean space X, is not a boolean product of subalgebras of $[0, 1]$.

6.6 Completeness

By definition, a lattice L is *complete* iff every subset $\{x_i \mid i \in I\}$ of L has a supremum and an infimum, which we respectively denote by

$$\bigvee_{i \in I} x_i, \quad (\text{or} \quad \bigvee\{x_i \mid i \in I\})$$

and

$$\bigwedge_{i \in I} x_i, \quad (\text{or} \quad \bigwedge\{x_i \mid i \in I\}).$$

Any complete lattice has a minimum element $0 = \bigvee \emptyset$ and a maximum element $1 = \bigwedge \emptyset$.

Definition 6.6.1 We say that an MV-algebra A is *complete* iff its underlying lattice $\mathbf{L}(A)$ is complete. We say that A is *σ-complete* iff suprema and infima exist for all finite or denumerable subsets in $\mathbf{L}(A)$.

By Theorem 3.5.1, the only complete and simple MV-algebras are $[0, 1]$ and the finite chains Ł_n.

Proposition 6.6.2 *Every σ-complete MV-algebra A (whence, a fortiori, every complete MV-algebra) is semisimple.*

Proof: Assuming $x \in Rad(A)$, we shall show that $x = 0$. As a matter of fact, by Proposition 3.6.4 we have

(6.6) $\quad n.x \leq \neg x, \quad$ i.e., $\quad x \odot n.x = 0.$

Let $s = \bigvee_{n \in \mathbf{N}}(n.x)$. Then for all $n = 0, 1, \ldots,$ we have $(n + 1).x \leq s$. From the monotonicity of the $\odot$ operation it follows that $(n+1).x \ominus x \leq s \ominus x$. From (6.6) we now get

$$n.x = n.x \wedge \neg x = (n.x \oplus x) \ominus x = (n + 1).x \ominus x \leq s \ominus x.$$

Therefore, $s \leq s \ominus x \leq s$, whence, $0 = s \ominus (s \ominus x) = s \wedge x = x$, as required. □

Lemma 6.6.3 *Let A be a complete MV-algebra. Let $\{x_i \mid i \in I\} \subseteq A$. Then*

$$(6.7) \quad \bigwedge_{i\in I} x_i = \neg \bigvee_{i\in I} \neg x_i$$

and

$$(6.8) \quad \bigvee_{i\in I} x_i = \neg \bigwedge_{i\in I} \neg x_i\,.$$

Proof: From the elementary properties of the natural order relation in A. □

Lemma 6.6.4 *Let A be a complete MV-algebra. Let $\{x_i \mid i \in I\} \subseteq A$. Then for each $x \in A$, the following* generalized distributive laws *hold:*

$$(6.9) \quad x \wedge \bigvee_{i\in I} x_i = \bigvee_{i\in I} (x \wedge x_i)$$

and

$$(6.10) \quad x \vee \bigwedge_{i\in I} x_i = \bigwedge_{i\in I} (x \vee x_i).$$

Proof: We first prove

$$(6.11) \quad x \odot \bigvee_{i\in I} x_i = \bigvee_{i\in I} (x \odot x_i)\,.$$

To this purpose, let $a = \bigvee_{i\in I} x_i$. Since $x_i \leq a$, for every $i \in I$, $x \odot x_i \leq x \odot a$. Let us assume that for each $i \in I$, $x \odot x_i \leq z$. By Lemma 1.1.4(iii), $x_i \leq \neg x \oplus z$ for each $i \in I$. Hence $a \leq \neg x \oplus z$, and by the same lemma, we get $x \odot a \leq z$. Thus, $x \odot a = \bigvee_{i\in I}(x \odot x_i)$, which settles (6.11). Now by (6.8), for every $i \in I$ we can write $\neg a \leq \neg x_i$; hence, by (6.11)

$$x \wedge a = (\neg a \oplus x) \odot a = \bigvee_{i\in I} ((\neg a \oplus x) \odot x_i)$$

$$\leq \bigvee_{i\in I}((\neg x_i \oplus x) \odot x_i) = \bigvee_{i\in I}(x \wedge x_i).$$

From the trivial inequality $\bigvee_{i\in I}(x \wedge x_i) \leq x \wedge a$ we finally obtain (6.9). To complete the proof, it is sufficient to note that, by Lemma 6.6.3, the identities (6.9) and (6.10) are equivalent. □

Corollary 6.6.5 *Let A be a complete MV-algebra. Then*

(i) $\mathbf{B}(A)$ *is a complete boolean algebra. As a matter of fact, for every set* $\{b_i \mid i \in I\} \subseteq \mathbf{B}(A)$ *we have*

(6.12) $\bigvee_{i\in I} b_i \in \mathbf{B}(A)$

and

(6.13) $\bigwedge_{i\in I} b_i \in \mathbf{B}(A)$;

(ii) For every $b \in \mathbf{B}(A)$, *letting* $(b]$ *and* $h_b : A \to (b]$ *be as in (6.1) and (6.4), it follows that* $(b]$ *is a complete MV-algebra, and* h_b *preserves arbitrary infima and suprema. In more detail,*

(6.14) $h_b\left(\bigvee_{i\in I} x_i\right) = \bigvee_{i\in I} h_b(x_i)$

and

(6.15) $h_b\left(\bigwedge_{i\in I} x_i\right) = \bigwedge_{i\in I} h_b(x_i)$.

Proof: (i) Let $x = \bigvee_{i\in I} b_i$. By (6.7) and (6.9), we have

$$\neg x = \bigwedge_{i\in I} \neg b_i \quad\text{and}\quad \neg x \wedge x = \bigvee_{i\in I}(\neg x \wedge b_i) \;\leq\; \bigvee_{i\in I}(\neg b_i \wedge b_i) = 0.$$

Then by Theorem 1.5.3(iii), x is a boolean element of A; this settles (6.12). The proof of (6.13) is similar.

(ii) Since $(b]$ is a complete MV-algebra, (6.14) and (6.15) are an immediate consequence of the definition of h_b, in the light of (6.9) and (6.10). □

Using Lemma 6.6.4 an easy adaptation of the proof of Lemma 6.4.5 yields

Lemma 6.6.6 *Let A be a complete MV-algebra. Let the set $\{b_i \mid i \in I\} \subseteq \mathbf{B}(A)$ satisfy the following conditions: $\bigvee_{i\in I} b_i = 1$, and $b_i \wedge b_j = 0$ whenever $i \neq j$. Then A is isomorphic to the direct product MV-algebra $\prod\{(b_i] \mid i \in I\}$.* □

From Lemmas 6.4.4 and 6.6.4 we get that for every complete MV-algebra A, there is a one-one correspondence between the direct product decompositions of A and the sets $\{b_i \mid i \in I\} \subseteq \mathbf{B}(A)$ such that $\bigvee_{i\in I} b_i = 1$ and $b_i \wedge b_j = 0$ whenever $i \neq j$. More generally, given an infinite cardinal κ and a κ-complete MV-algebra A (as defined by a natural extension of 6.6.1), an easy adaptation of the proofs of Lemmas 6.4.4 and 6.6.4 yields a one-one correspondence between direct product decompositions of A into κ many factors, and sets $\{b_\alpha \mid \alpha \in \kappa\} \subseteq \mathbf{B}(A)$ such that $\bigvee_{\alpha\in\kappa} b_\alpha = 1$ and $b_\alpha \wedge b_\beta = 0$ whenever $\alpha \neq \beta$. The crucial point is that in Lemma 6.6.4 the supremum on the right hand side exists iff so does the supremum on the left hand side, and if this is the case the suprema coincide.

Lemma 6.6.7 *Let $\{A_i \mid i \in I\}$ be a family of MV-algebras. Then the product $P = \prod_{i\in I} A_i$ is a complete MV-algebra iff so is each A_i.*

Proof: It is easy to check that the two lattices $\mathbf{L}(P)$ and $\prod_{i\in I} \mathbf{L}(A_i)$ coincide. Clearly, the direct product of complete lattices is a complete lattice. Thus, if each A_i is complete, then so is P. The converse is an immediate consequence of Corollary 6.6.5(ii), together with Lemma 6.4.4. □

6.7 Atoms and Pseudocomplements

Definition 6.7.1 By an *atom* of an MV-algebra A we mean an atom of the underlying lattice $\mathbf{L}(A)$. We say that A is *atomic* iff for each $0 \neq x \in A$ there is an atom $a \in A$ with $a \leq x$. We say that A is *atomless* iff no element of A is an atom.

Examples. For each $n \geq 2$, the element $1/(n-1)$ is an atom of the MV-chain $Ł_n$. By Proposition 3.5.3, all infinite subalgebras of $[0, 1]$ are atomless. By Theorem 3.5.1 and Proposition 3.5.3, up to isomorphism, simple, complete, atomic MV-algebras coincide with the MV-chains $Ł_n$, where $n \geq 2$. Further, $[0,1]$ is the only simple, complete and atomless MV-algebra.

In the rest of this section, we shall study algebras of the form $(\,b\,]$, when b is an atom of the boolean algebra $\mathbf{B}(A)$.

Let A be a complete MV-algebra. For each $z \in A$, recall that, by definition, $J_z{}^{\perp} = \{x \in A \mid x \wedge z = 0\}$. It is not hard to see that $J_z{}^{\perp}$ is an ideal of A. (To this purpose, using the subdirect representation theorem one simply notes that for all $p, q, r \in A$ if $p \wedge q = 0$ then $p \wedge r = p \wedge (q \oplus r)$).

Let $z^* \in A$ be defined by $z^* = \bigvee J_z{}^{\perp} = \bigvee\{x \mid x \wedge z = 0\}$. Then z^* is the *pseudocomplement* of z, in the sense that for any $x \in A$

(6.16) $x \wedge z = 0$ iff $x \leq z^*$.

Lemma 6.7.2 *Let A be a complete MV-algebra. Then the lattice $\mathbf{L}(A)$ is* pseudocomplemented, *in the sense that each $z \in A$ has its pseudocomplement z^*. Moreover, for each $z \in A$ the pseudocomplement z^* is a boolean element of A.*

Proof: By Lemma 6.6.4, $\mathbf{L}(A)$ is pseudocomplemented. By the above discussion, together with the assumed completeness of A and Lemma 6.6.4(4), for every element $z \in A$, the ideal $J_z{}^{\perp}$ coincides with $(z^*]$. By Corollary 1.5.6, $z^* \in \mathbf{B}(A)$. □

With the above notation, a straightforward computation yields

(6.17) $x \leq x^{**}$;

(6.18) If $x \leq y$ then $x^{**} \leq y^{**}$;

(6.19) $(x \wedge y)^{**} = x^{**} \wedge y^{**}$;

(6.20) If $z \in \mathbf{B}(L)$ then $z^{**} = z$.

Theorem 6.7.3 *Let A be a complete MV-algebra and z an atom of $\mathbf{B}(A)$. If there is an atom a of A such that $a \leq z$, then the MV-algebra $(\,z\,]$ is isomorphic to the finite chain $Ł_n$, for some $n \geq 2$. If no such atom exists, then $(\,z\,]$ is isomorphic to $[0,1]$.*

Proof: We first prove that $(\neg z]$ is a prime ideal in $\mathbf{L}(A)$. Let us assume $x \wedge y \leq \neg z$. By (6.17)-(6.20),

$$x^{**} \wedge y^{**} = (x \wedge y)^{**} \leq (\neg z)^{**} = \neg z.$$

Since, by Lemma 6.7.2, both x^{**} and y^{**} are elements of $\mathbf{B}(A)$, we must either have $x^{**} \leq \neg z$ or $y^{**} \leq \neg z$. By (6.17), either $x \leq \neg z$ or $y \leq \neg z$. Therefore, $(\neg z]$ is a prime ideal of $\mathbf{L}(A)$ and, by Corollary 1.5.6, it is also a prime ideal of A. Thus the quotient MV-algebra $A/(\neg z]$ is totally ordered. By Corollary 6.4.1, $A/(\neg z]$ is isomorphic to $(z]$; by Corollary 6.6.5(ii), $(z]$ is complete. Since a is an atom of $(z]$ iff a is an atom of A and $a \leq z$, the desired result now follows from the observation that $[0,1]$ is the only simple complete atomless MV-algebra. □

Lemma 6.7.4 *For each atom a of a complete MV-algebra A, a^{**} is an atom of the boolean algebra $\mathbf{B}(A)$.*

Proof: By Lemma 6.7.2, $a^{**} \in \mathbf{B}(A)$ and, by (6.18), $a^{**} \neq 0$. Let us assume that $z \in \mathbf{B}(A)$ and $z \leq a^{**}$. By (6.19) and (6.20),

$$z = z \wedge a^{**} = z^{**} \wedge a^{**} = (z \wedge a)^{**},$$

and since a is an atom of A, we must either have $z \wedge a = 0$ or $z \wedge a = a$. Therefore, $z = 0$ or $z = a^{**}$. □

6.8 Complete distributivity

As usual, a complete lattice L is said to be *completely distributive* iff for every family $\{J_i \mid i \in I\}$ of nonempty sets, and for arbitrary $x_{ij} \in L$, letting $T = \prod_{i \in I} J_i$ we have

$$\bigwedge_{i \in I} \bigvee_{j \in J_i} x_{ij} = \bigvee_{f \in T} \bigwedge_{i \in I} x_{if(i)} \quad \text{and} \quad \bigvee_{i \in I} \bigwedge_{j \in J_i} x_{ij} = \bigwedge_{f \in T} \bigvee_{i \in I} x_{if(i)}\,.$$

We say that a complete MV-algebra A is *completely distributive* iff its underlying lattice $\mathbf{L}(A)$ is completely distributive in the above sense.

Theorem 6.8.1 *For every MV-algebra A the following conditions are equivalent:*
(i) A is a direct product of totally ordered complete MV-algebras;
(ii) A is complete and completely distributive;
(iii) A is complete and the boolean algebra $\mathbf{B}(A)$ *is atomic.*

Proof: (i) $\Rightarrow$ (ii): Trivial.

(ii) $\Rightarrow$ (iii): By Corollary 6.6.5, the boolean algebra $\mathbf{B}(A)$ is complete and completely distributive. Then, by a classical result of Tarski, $\mathbf{B}(A)$ is isomorphic to a powerset boolean algebra. Thus in particular, $\mathbf{B}(A)$ is atomic.

(iii) $\Rightarrow$ (i): The atoms of $\mathbf{B}(A)$ form a set $\{z_i \mid i \in I\}$ such that $z_i \wedge z_j = 0$ whenever $i \neq j$; further, $\bigvee_{i \in I} z_i = 1$. By Lemma 6.6.6, A is isomorphic to the product MV-algebra $\prod_{i \in I}(\, z_i\,]$. By Theorem 6.7.3, each $(\, z_i\,]$ is a complete chain. $\Box$

Corollary 6.8.2 *An MV-algebra is the direct product of copies of* $[0, 1]$ *iff it is complete, completely distributive, and atomless.* $\Box$

Corollary 6.8.3 *An MV-algebra is a direct product of finite chains iff it is complete and atomic.* $\Box$

As a corollary, we have a new proof of the result (Proposition 3.6.5) that every finite MV-algebra is a direct product of finite chains.

Definition 6.8.4 An MV-algebra A is said to *have no linear factors* iff whenever A can be written as a direct product of MV-algebras,

$$A = \prod_{i \in I} A_i,$$

then no A_i is totally ordered. For any complete MV-algebra A let $\{a_i \mid i \in I\}$ be the set of atoms of A. Let $\{z_j \mid j \in J\}$ be the set of atoms of $\mathbf{B}(A)$. Then we define

$$s_A = \bigvee_{i \in I} a_i^{**}, \quad \text{and} \quad z_A = \bigvee_{i \in I} z_i.$$

In the rest of this section, whenever in a product MV-algebra one of the factors $(x]$ coincides with the singleton $(0]$, we shall tacitly understand that $(x]$ must be deleted.

Theorem 6.8.5 *If A is a complete MV-algebra then*

(6.21) $A = (s_A] \times (\neg s_A \wedge z_A] \times (\neg z_A]$.

If $s_A > 0$ then $(s_A]$ is a direct product of finite chains. If $s_A \neq z_A$ then $(\neg s_A \wedge z_A]$ is a direct product of copies of $[0,1]$, and the MV-algebra $(\neg z_A]$ is complete and has no linear factors.

Proof: Formula (6.21) immediately follows from Lemmas 6.6.6 and 6.7.4. If $0 \neq z \in \mathbf{B}(\neg z_A]$, then z is not an atom of $\mathbf{B}(A)$ and, by Proposition 6.4.3, $(z]$ is not totally ordered. By Lemma 6.4.4, $(\neg z_A]$ has no linear factors. Assume $s_A \neq 0$ and let $x \in (s_A], x \neq 0$. If $a_i \not\leq x$ for each $i \in I$ we obtain

$$x = x \wedge s_A = x \wedge \bigvee_{i\in I} a_i^{**} \leq \bigvee_{i\in I}(x^{**} \wedge a_i^{**}) = \bigvee_{i\in I}(x \wedge a_i)^{**} = 0,$$

which is impossible. Therefore, $a_i \leq x$, for some $i \in I$, whence $(s_A]$ is a complete atomic MV-algebra. By Corollary 6.8.3, $(s_A]$ is a direct product of finite chains. By Lemma 6.7.4, $s_A \leq z_A$. Thus if $s_A \neq z_A$, then $\neg s_A \wedge z_A \neq 0$. If a were an atom of $(\neg s_A \wedge z_A]$, then a would also be an atom of A; then $a \leq a^{**} < s_A$, whence $a = 0$, which is impossible. Therefore, $(\neg s_A \wedge z_A]$ is an atomless complete MV-algebra. The set of atoms of $\mathbf{B}((\neg s_A \wedge z_A]) = (\neg s_A \wedge z_A] \cap \mathbf{B}(A)$ coincides with the set of those atoms z_i of $\mathbf{B}(A)$ such that $z_i \leq \neg s_A$. From the identities

$$\neg s_A \wedge z_A = \neg s_A \wedge \bigvee_{i\in I} z_i = \bigvee_{i\in I}(\neg s_A \wedge z_i)$$

it follows that $\mathbf{B}((\neg s_A \wedge z_A])$ is an atomic boolean algebra. By Theorem 6.8.1 and Corollary 6.8.2, if $s_A \neq z_A$ then $(\neg s_A \wedge z_A]$ is a direct product of copies of $[0,1]$. □

6.9 Bibliographical remarks

In every MV-algebra A the operation $\neg$ satisfies the following conditions:

(M1) $\neg 0 = 1$;

(M2) $\neg\neg x = x$;

(M3) $\neg(x \vee y) = \neg x \wedge \neg y$.

A distributive lattice with 0 and 1 equipped with an operation satisfying (M1), (M2) and (M3), is called *De Morgan algebra*. A is called a *Kleene algebra* (see [12]) iff it satisfies the additional condition:

(K) $x \wedge \neg x \ \leq \ y \vee \neg y$.

By the subdirect representation theorem, for every MV-algebra A, its underlying lattice $(A, \vee, \wedge, \neg, 0, 1)$ is a Kleene algebra.

Theorem 6.1.3 was first proved in [146], and was used in [99] and [100] to show that, in Zermelo-Fraenkel set theory without the axiom of choice, the following statements are equivalent (one implication is trivial):

- Every MV-algebra has a maximal ideal.
- Every boolean algebra has a maximal ideal.

It is known that this latter statement is strictly weaker than the axiom of choice.

The result proved in the claim in Theorem 6.1.5 is a well known elementary fact of the theory of distributive lattices, due to Birkhoff and Stone. See, e.g., [103]. For a study of stonean ideals in lattices see [45]. The proof of the claim in Proposition 6.3.3 can be found in [163, Lemma 8.5]. Proposition 6.4.1 is due to Rodríguez, see [212]. For direct product decompositions also see [123].

For boolean products of MV-algebras see [231], [232], [55] and [56]. Corollary 6.5.6 is due to Torrens [231] (caution: our hyperarchimedean MV-algebras are called archimedean in [231]).

Proposition 6.6.2 is due to [132]. In the same paper the author also proved Lemma 6.6.4. Also see [49]. The direct algebraic proof presented here is due to [14]. Equations (6.17)-(6.20) hold in every distributive pseudocomplemented lattice (see [103], or [12]).

For a proof of the classical result of Tarski referred to in the proof of Theorem 6.8.1 see, for instance, [25], Chapter 5, §5, Theorem 17. Corollary 6.8.2 is due to Bosbach (see [31]). In the same paper one can also find a proof of Corollary 6.8.3 (see also [49]).

Complete MV-algebras, convergence properties, various kinds of topological and order completions are considered, e.g., in [91], [89], [124], [14], [15], [128], [125], [126], [69], [224], [119] and [127]. For model-completions see [135] and [136].

Chapter 7

MV-algebras and ℓ-groups

As proved at the beginning of Chapter 2, $\mathbf{\Gamma}$ is a functor from the category $\mathcal{A}$ of ℓ-groups with a distinguished strong unit, to the category $\mathcal{MV}$ of MV-algebras. In this chapter we shall prove that $\mathbf{\Gamma}$ is a natural equivalence (i.e., a full, faithful and dense functor) between $\mathcal{A}$ and $\mathcal{MV}$. As a consequence, a genuine addition can be uniquely recovered from the MV-algebraic structure. Several applications will be discussed.

7.1 Inverting the functor Γ

In this section we shall give an explicit construction of an adjoint functor of $\mathbf{\Gamma}$. Our starting point is the ℓ-group G_A with order unit u_A considered in Section 2 of Chapter 2. As the reader will recall, for every MV-algebra A, G_A is an ℓ-group and there is an isomorphism between the ordered monoid ${G_A}^+$ of its positive elements and the ordered monoid M_A of good sequences of A.

Let A and B be MV-algebras, and $h : A \to B$ a homomorphism. If $\mathbf{a} = (a_1, a_2, \ldots)$ is a good sequence of A, then $(h(a_1), h(a_2), \ldots)$ is a good sequence of B. Let $h^* : M_A \to M_B$ be defined by

(7.1) $h^*(\mathbf{a}) = (h(a_1), h(a_2), \ldots)$,

for all $\mathbf{a} \in M_A$. By direct inspection, using (2.11) and Proposition 2.2.3, for all $\mathbf{a}, \mathbf{b} \in M_A$ we have

(7.2) $h^*(\mathbf{a} + \mathbf{b}) = h^*(\mathbf{a}) + h^*(\mathbf{b})$;

(7.3) $h^*(\mathbf{a} \vee \mathbf{b}) = h^*(\mathbf{a}) \vee h^*(\mathbf{b})$;

(7.4) $h^*(\mathbf{a} \wedge \mathbf{b}) = h^*(\mathbf{a}) \wedge h^*(\mathbf{b})$.

Thus, $h^* : M_A \to M_B$ is both a monoid-homomorphism and a lattice-homomorphism. Let us further define the map $h^\# : G_A \to G_B$ by

(7.5) $h^\#([\mathbf{a}, \mathbf{b}]) = [h^*(\mathbf{a}), h^*(\mathbf{b})]$.

Let u_A and u_B be the strong units of G_A and G_B given by Proposition 2.4.4. Then by (7.2)-(7.4), the map $h^\#$ is a unital ℓ-group homomorphism of $\langle G_A, u_A\rangle$ into $\langle G_B, u_B\rangle$. Let us agree to write

(7.6) $\Xi(A) = \langle G_A, u_A\rangle$ and $\Xi(h) = h^\#$.

Proposition 7.1.1 *Ξ is a functor from $\mathcal{MV}$ into $\mathcal{A}$.* □

In our present notation, Theorem 2.4.5 states that the map $a \mapsto \varphi_A(a) = [(a), (0)]$ defines an isomorphism of the MV-algebras A and $\Gamma(\Xi(A))$. Using the maps φ_A ($A \in \mathcal{MV}$), we obtain

Theorem 7.1.2 *The composite functor $\Gamma\Xi$ is* naturally equivalent *to the identity functor of $\mathcal{MV}$. In other words, for all MV-algebras A, B and homomorphism $h : A \to B$, we have a commutative diagram*

$$\begin{array}{ccc} A & \xrightarrow{h} & B \\ \varphi_A \downarrow & & \downarrow \varphi_B \\ \Gamma(\Xi(A)) & \xrightarrow{\Gamma(\Xi(h))} & \Gamma(\Xi(B)) \end{array}$$

in the sense that, for each $a \in A$, $\varphi_B(h(a)) = (\Gamma(\Xi(h)))(\varphi_A(a))$.

Proof: For each $a \in A$, $\varphi_B(h(a)) = [(h(a)), (0)]$ and $\varphi_A(a) = [(a), (0)]$. Further, by (7.1)-(7.6) $\Xi(h)([(a), (0)]) = [(h(a)), (0)]$, the latter being an element of $\Gamma(\Xi(B))$. Since $\Gamma(\Xi(h))$ is the restriction of $\Xi(h)$ to $\Gamma(\Xi(B))$, we can write $\Gamma(\Xi(h))(\varphi_A(a)) = [(h(a)), (0)] = \varphi_B(h(a))$, as required. □

Our next aim is to prove that the composite functor $\Xi\Gamma$ is also naturally equivalent to the identity functor of the category $\mathcal{A}$. We first prepare the "dual" of the maps φ_A.

Lemma 7.1.3 *Suppose G is an ℓ-group with order unit u, and let $A = \Gamma(G,u) \subseteq G$. For each $0 \leq a \in G$ there is a* unique *good sequence $g(a) = (a_1, \ldots, a_n)$ of elements of A such that $a = a_1 + \ldots + a_n$.*

Proof: For every integer $k \geq 1$, let a_k be inductively defined by: $a_1 = a \wedge u$, and $a_{k+1} = (a - a_1 - \cdots - a_k) \wedge u$.

In order to prove that $(a_1, \ldots, a_k, \ldots)$ is a good sequence of A, we first prove the identity

(7.7) $a - a_1 - \cdots - a_k = (a - ku) \vee 0 = (a - ku)^+$.

The proof is by induction on k. For $k = 1$, we have

$$a - a_1 = a - (a \wedge u) = 0 \vee (a - u) = (a - u)^+.$$

For the induction step, assume the identity to be true for $k = m$. Then

$$a - a_1 - \cdots - a_m - a_{m+1}$$

$$= (a - mu)^+ - ((a - mu)^+ \wedge u)$$

$$= 0 \vee ((a - mu)^+ - u)$$

$$= (a - (m + 1)u)^+,$$

which settles (7.7). Since for each $k \geq 1$, $a_k = (a - (k - 1)u)^+ \wedge u$, then necessarily each a_k is in the unit interval $[0, u]$ of G. Since u is a strong unit of G, there is an integer $n \geq 1$ such that $a \leq nu$, and hence, $a_j = 0$ for all $j > n$.

We shall now prove that $a_k \oplus a_{k+1} = a_k$. To simplify the notation, let $t = a - ku$. Then by (7.7)

$$a_k \oplus a_{k+1}$$

$$= ((t^+ \wedge u) + ((t - u)^+ \wedge u)) \wedge u$$

$$= (t^+ + (t - u)^+) \wedge u$$

$$= ((t \vee 0) + ((t - u) \vee 0)) \wedge u$$

$$= ((2t - u) \wedge u) \vee (t^+ \wedge u) = t^+ \wedge u = a_k.$$

We have used the inequality $t^+ \geq ((2t - u) \wedge u)$, which follows from

$$(t \vee 0) - ((2t - u) \wedge u)$$

$$= (u - t) \vee (t - u) \vee (u - 2t) \vee -u$$

$$\geq (u - t) \vee (t - u) = |t - u| \geq 0.$$

Hence $(a_1, \ldots, a_n)$ is a good sequence of elements of $A = [0, u]$, and

$$a - \sum_{i=1}^{n} a_i = a_{n+1} = a_{n+2} = \ldots = 0.$$

Finally, to prove uniqueness, note that if $(b_1, \ldots, b_m)$ is a good sequence of elements of $[0, u]$ such that $a = \sum_{i=1}^{m} b_i$, then by definition of good sequences together with Lemma 2.1.3(ii), we have

$$b_1 = b_1 \oplus \cdots \oplus b_m = (b_1 + \cdots + b_m) \wedge u = a \wedge u = a_1,$$

and, proceeding inductively, for $i \geq 2$,

$$b_i = b_i \oplus \cdots \oplus b_m$$

$$= (b_i + \cdots + b_m) \wedge u$$

$$= (a - b_1 - \cdots - b_{i-1}) \wedge u$$

$$= (a - a_1 - \cdots - a_{i-1}) \wedge u = a_i,$$

as required. □

It follows from the above lemma that the correspondence $a \mapsto g(a)$ defines an injective mapping from the positive cone G^+ of G, onto the monoid $M_{\Gamma(G,u)}$ of good sequences of $\Gamma(G, u)$. In order to show that this mapping is both a monoid-isomorphism and a lattice-isomorphism, we prepare the following

Definition 7.1.4 An *ℓ-ideal* of an ℓ-group G is a subgroup J of G satisfying the following condition:

(7.8) If $x \in J$ and $|y| \leq |x|$ then $y \in J$.

Letting for each $a \in G$, $a/J = a + J$ be the coset determined by a, and defining $a/J \vee b/J = (a \vee b)/J$, the quotient group G/J becomes an ℓ-group. The map $q_J : G \to G/J$ sending each $a \in G$ to the coset a/J, is an ℓ-homomorphism such that $Ker(q_J) = J$. Moreover, if u is a strong unit of G, then $q_J(u)$ is a strong unit of G/J. Conversely, if G and H are ℓ-groups and $f: G \to H$ is an ℓ-homomorphism, then $Ker(f) = f^{-1}(\{0\})$ is an ℓ-ideal of G, and $G/Ker(f)$ is isomorphic to the ℓ-subgroup $f(G)$ of H.

An ℓ-ideal J of an ℓ-group G is said to be *prime* iff J is proper (i.e., $J \neq G$), and the quotient ℓ-group G/J is totally ordered.

Lemma 7.1.5 *Let G be an ℓ-group with an order unit u. Then the mapping $g: G^+ \to M_{\Gamma(G,u)}$ of Lemma 7.1.3 satisfies the following conditions, for all $a, b \in G^+$:*

(i) $g(a+b) = g(a) + g(b)$;

(ii) $g(a \vee b) = g(a) \vee g(b)$;

(iii) $g(a \wedge b) = g(a) \wedge g(b)$;

(iv) $g(u) = (u)$.

Proof: The notion of a subdirect product of ℓ-groups is, *mutatis mutandis*, the same as the corresponding notion for MV-algebras (Definition 1.3.1). The following claim is a classical result of Birkhoff :

Claim. Every ℓ-group G is a subdirect product of totally ordered abelian groups.

Indeed, let $0 \neq a \in G$. An adaptation of the proof of Proposition 1.2.13 shows that any ℓ-ideal of G which is maximal for not containing a is automatically prime. It follows that the intersection of all prime ℓ-ideals of G is the zero ideal, and G can be embedded into the product of all its prime quotients G/J. Thus, G is a subdirect product of totally ordered ℓ-groups, and our claim is proved.

Let X be the set of prime ℓ-ideals of G. Our claim yields a family $\{G_x\}_{x\in X}$ of totally ordered abelian groups and a one-one ℓ-homomorphism

$$h: G \to \prod_{x\in X} G_x$$

such that, for each $x \in X$, the composite function $\pi_x h$ maps G onto G_x (π_x being the canonical projection onto G_x). For every $a \in G$, letting $a_x = \pi_x(h(a))$, it follows that u_x is a strong unit of G_x. Given arbitrary elements $a, b \in G^+$, let us write $g(a) = (a_1, \ldots, a_m)$ and $g(b) = (b_1, \ldots, b_n)$. Then, for each $x \in X$, both $(a_{1x}, \ldots, a_{mx})$ and $(b_{1x}, \ldots, b_{nx})$ are good sequences in $\Gamma(G_x, u_x)$. Moreover,

$$a_x = \sum_{i=1}^{m} a_{ix} \quad \text{and} \quad b_x = \sum_{i=1}^{n} b_{ix}.$$

Since the MV-algebra $\Gamma(G_x, u_x)$ is totally ordered, by Proposition 2.2.1 every good sequence of A has the form $(1^p, a)$. Thus, there are elements $\alpha_x, \beta_x \in G_x$ with $0 \leq \alpha_x < u_x$ and $0 \leq \beta_x < u_x$, satisfying the equations $(a_{1x}, \ldots, a_{mx}) = (u_x^{p_x}, \alpha_x)$ and $(b_{1x}, \ldots, b_{nx}) = (u_x^{q_x}, \beta_x)$. Further, by definition of g,

$$(7.9) \qquad a_x = p_x u_x + \alpha_x \quad \text{and} \quad b_x = q_x u_x + \beta_x,$$

whence $(a+b)_x = a_x + b_x = (p_x + q_x)u_x + \alpha_x + \beta_x$. Recalling (2.12), by an application of Lemma 2.1.3(i), for all $x \in X$ we obtain

$$(u_x^{p_x}, \alpha_x) + (u_x^{q_x}, \beta_x) = (u_x^{p_x+q_x}, \alpha_x \oplus \beta_x, \alpha_x \odot \beta_x).$$

Then, by direct inspection, $(a+b)_x = (a_{1x}, \ldots, a_{mx}) + (b_{1x}, \ldots, b_{nx})$. We conclude that $g(a+b) = g(a) + g(b)$, and (i) is proved.

To prove (ii), for any $x \in X$, we either have $a_x \leq b_x$ or $b_x \leq a_x$. Suppose, without loss of generality, $a_x \leq b_x$. Then, from $0 \leq \alpha_x \leq \beta_x < u_x$, by (7.9) we get $p_x \leq q_x$, whence $(u_x^{p_x}, \alpha_x) \leq (u_x^{q_x}, \beta_x)$. Therefore, letting $r = \max(m, n)$,

$$(a \vee b)_x = \sum_{i=1}^{r} (a_{ix} \vee b_{ix}).$$

In other words, $a \vee b = \sum_{i=1}^{r}(a_i \vee b_i)$, whence, by Lemma 7.1.3, $g(a \vee b) = g(a) \vee g(b)$, which completes the proof of (ii). The proof of (iii) is similar, and (iv) follows at once by definition of g. □

Corollary 7.1.6 *For every $\langle G, u\rangle \in \mathcal{A}$ let the map $\psi_{\langle G,u\rangle} : G \to G_{\Gamma(G,u)}$ be defined by*

$$(7.10) \quad \psi_{\langle G,u\rangle}(a) = [g(a^+), g(a^-)],$$

for all $a \in G$. It follows that $\psi_{\langle G,u\rangle}$ is an ℓ-group isomorphism of G onto $G_{\Gamma(G,u)}$, and $\psi_{\langle G,u\rangle}(u) = [(u), (0)]$. □

In the light of Corollary 7.1.6, using the maps $\psi_{\langle G,u\rangle}$ (for all $\langle G, u\rangle \in \mathcal{A}$), we have the following

Theorem 7.1.7 *The composite functor $\Xi\Gamma$ is* naturally equivalent *to the identity functor of the category $\mathcal{A}$. In other words, for any two ℓ-groups with strong unit $\langle G, u\rangle$ and $\langle H, v\rangle$ and unital ℓ-homomorphism $f : \langle G, u\rangle \to \langle H, v\rangle$, we have a commutative diagram*

$$\begin{array}{ccc} \langle G,u\rangle & \xrightarrow{f} & \langle H,v\rangle \\ \psi_{\langle G,u\rangle}\downarrow & & \downarrow\psi_{\langle H,v\rangle} \\ \Xi(\Gamma(G,u)) & \xrightarrow{\Xi(\Gamma(f))} & \Xi(\Gamma(H,v)) \end{array}$$

in the sense that, for all $a \in G$, $\psi_{\langle H,v\rangle}(f(a)) = \Xi(\Gamma(f))(\psi_{\langle G,u\rangle}(a))$.

Proof: By Lemma 7.1.3 we can write $g(a^+) = (a_1, \ldots, a_n)$, for a uniquely determined good sequence $(a_1, \ldots, a_n) \in M_{\Gamma(G,u)}$. Letting $h = \Gamma(f)$, we then obtain

$$f(a)^+ = f(a^+) = \sum_{i=1}^{n} f(a_i) = \sum_{i=1}^{n} h(a_i),$$

whence, recalling (7.1), $g(f(a)^+) = (h(a_1), \ldots, h(a_n)) = h^*(g(a^+))$. Similarly, $g(f(a)^-) = h^*(g(a^-))$, whence by (7.5), (7.6) and (7.10),

$$\psi_{\langle H,v\rangle}(f(a))$$

$$= [g(f(a)^+), g(f(a)^-)]$$

$$= [h^*(g(a^+)), h^*(g(a^-))]$$

$$= h^\#([g(a^+), g(a^-)])$$

$$= \Xi(\Gamma(f))([g(a^+), g(a^-)])$$

$$= \Xi(\Gamma(f))(\psi_{\langle G,u\rangle}(a)).$$

□

From Theorems 7.1.2 and 7.1.7 we immediately get

Corollary 7.1.8 *The Γ functor defines a natural equivalence between the category $\mathcal{A}$ of ℓ-groups with strong unit, and the category $\mathcal{MV}$ of MV-algebras.* □

7.2 Applications

Lemma 7.2.1 *Let $h: \Gamma(G, u) \to \Gamma(H, v)$ be a homomorphism of MV-algebras.*

(i) There is a unique *unital ℓ-homomorphism $f: \langle G, u\rangle \to \langle H, v\rangle$ such that $h = \Gamma(f)$;*

(ii) If h maps $\Gamma(G, u)$ onto $\Gamma(H, v)$ then f maps G onto H;

(iii) If h maps $\Gamma(G, u)$ one-one into $\Gamma(H, v)$ then f maps G one-one into H.

Proof: (i) In the light of Corollary 7.1.6 and Theorem 7.1.7, let f be defined by

$$f = \psi^{-1}_{\langle H,v\rangle}(\Xi(h))\psi_{\langle G,u\rangle}.$$

By (7.5), (7.6) and (7.10), for each $a \in G$ we can write

$$f(a) = \psi^{-1}_{\langle H,v\rangle}([h^*(g(a^+)), h^*(g(a^-))]).$$

In the particular case when $a \in [0, u]$, since $g(a^+) = g(a) = (a)$ then $[h^*(g(a^+)), h^*(g(a^-))] = [(h(a)), (0)]$, whence

$$f(a) = \psi^{-1}_{\langle H,v\rangle}([(h(a)), (0)]) = h(a).$$

Therefore, $\Gamma(f) = h$. In order to establish the uniqueness of f, suppose $f(b) \neq f'(b)$ for some $b \in G^+$, and unital ℓ-homomorphism

$$f': \langle G, u\rangle \to \langle H, v\rangle.$$

Writing $b = a_1 + \ldots + a_k$ for suitable $a_i \in \Gamma(G,u)$, if $\Gamma(f) = \Gamma(f')$ we have $f(b) = f(a_1) + \ldots + f(a_k) = f'(a_1) + \ldots + f'(a_k) = f'(b)$, a contradiction.

(ii) Suppose that h maps $\Gamma(G,u)$ onto $\Gamma(H,v)$, and let $b \in \Gamma(H,v)$. Then there exists an element $a \in \Gamma(G,u)$ such that $h(a) = b$ and $f(a) = h(a) = b$. Since every element of H^+ (resp., every element of G^+) is a sum of elements of the unit interval $[0,v]$ (resp., a sum of elements of $[0,u]$) we have that f maps G^+ onto H^+, whence f maps G onto H.

(iii) Finally, if f is not one-one then f maps some nonzero element $x \in G$ into the zero element of H. Without loss of generality, $0 \leq x$, whence $0 = f(x \wedge u) = h(x \wedge u)$, and h is not one-one. □

Let us agree to denote by $\mathcal{I}(G)$ the set of all ℓ-ideals of G, ordered by inclusion.

Theorem 7.2.2 *Let G be an ℓ-group with strong unit u, and $A = \Gamma(G,u)$. Then the correspondence*

$$\phi : J \mapsto \phi(J) = \{x \in G \mid |x| \wedge u \in J\}$$

is an order-isomorphism from the set $\mathcal{I}(A)$ of ideals of the MV-algebra A, ordered by inclusion, onto $\mathcal{I}(G)$. The inverse isomorphism ψ is given by $H \in \mathcal{I}(G) \mapsto \psi(H) = H \cap [0,u]$.

Proof: Trivially, $\psi(H) \in \mathcal{I}(A)$ whenever $H \in \mathcal{I}(G)$. Conversely, to prove that for each ideal J of A, $\phi(J)$ is an ℓ-ideal of G, assume $x, y \in \phi(J)$. Then $(|x| \wedge u) \oplus (|y| \wedge u) \in J$. Furthermore,

$$(|x| \wedge u) \oplus (|y| \wedge u)$$

$$= ((|x| \wedge u) + (|y| \wedge u)) \wedge u$$

$$= ((|x| \wedge u) + |y|) \wedge ((|x| \wedge u) + u) \wedge u$$

$$= ((|x| \wedge u) + |y|) \wedge u$$

$$= (|x| + |y|) \wedge u$$

$$\geq |x - y| \wedge u,$$

whence $|x-y| \wedge u \in J$, and hence, $x - y \in \phi(J)$. Since $0 \in \phi(J)$, then $\phi(J)$ is a subgroup of G, and since $\phi(J)$ satisfies (7.8), $\phi(J)$ is an ℓ-ideal of G. Trivially, the function $\phi: \mathcal{I}(A) \to \mathcal{I}(G)$ preserves inclusions. Now it is not hard to see that $\psi(\phi(J)) = J$, for each ideal J of A. Hence to complete the proof, there remains to be shown that for each ℓ-ideal H of G, $\phi(\psi(H)) = H$; stated otherwise, $H = \{x \in G \mid |x| \wedge u \in H \cap [0,u]\}$. The inclusion $H \subseteq \{x \in G \mid |x| \wedge u \in H \cap [0,u]\}$ is clear from (7.8). On the other hand, if $|x| \wedge u \in H \cap [0,u]$, then since u is a strong unit there is an integer $n \geq 1$ such that $|x| \leq nu$. Thus, $0 \leq |x| = |x| \wedge nu \leq n(|x| \wedge u) \in H$, and $x \in H$. □

Corollary 7.2.3 *The map $J \mapsto J \cap [0,u]$ defines an isomorphism between the partially ordered set of prime ℓ-ideals of G and the partially ordered set of prime ideals of $\Gamma(G,u)$, both sets being equipped with the inclusion ordering.* □

Theorem 7.2.4 *Let G be an ℓ-group with strong unit u. Then for every ℓ-ideal J of G, we have the isomorphism*

$$\Gamma(G/J, u/J) \cong \Gamma(G,u)/(J \cap [0,u]).$$

Proof: Let $f: G \to G/J$ be the natural ℓ-homomorphism, and $k = \Gamma(f)$. Let $A = \Gamma(G,u)$ and $I = J \cap [0,u]$. By Theorem 7.2.2, I is an ideal of A. Since by Lemma 7.2.1(ii), k is a homomorphism of A onto $\Gamma(G/J, u/J)$, whose kernel coincides with I, then $A/I \cong \Gamma(G/J, u/J)$, as required. □

Remark: It follows at once that the correspondence $J \mapsto J \cap [0,1]$ defines a 1-1 mapping from the set of maximal ℓ-ideals of G onto the set of maximal ideals of $\Gamma(G,u)$.

We shall now construct uncountably many non-isomorphic simple MV-algebras: We first prepare the following well-known result; its proof is included here to increase readability.

Proposition 7.2.5 *Let G and H be ℓ-subgroups of the additive group $\mathbf{R}$ of real numbers with natural order. Assume $1 \in G \cap H$. Then there is at most one ℓ-isomorphism f of G onto H such that $f(1) = 1$. Whenever such f exists, then G necessarily coincides with H, and f is the identity function on G.*

Proof: By way of contradiction, assume $G \not\subseteq H$, and let $a \in G \setminus H$. Then $a \neq f(a) \in H$. In case $a < f(a)$ there is a rational number p/q such that $a < p/q < f(a)$, i.e., $qa < p < qf(a)$. Therefore, $0 < p - qa \in G$, and $0 > p - qf(a) = pf(1) - qf(a) = f(p - qa) \in H$, whence f does not preserve the order, a contradiction. In a similar way, if $f(a) < a$, a negative element of G would be mapped by f into a positive element of H. We have proved that $G \subseteq H$. Symmetrically, $H \subseteq G$, and we are done. □

Recalling now Lemma 7.2.1 we obtain

Corollary 7.2.6 *Two subalgebras A and B of $[0, 1]$ are isomorphic iff $A = B$; the identity function is the only automorphism of A.* □

Example: As in the example following Theorem 3.4.9, for any irrational number $\alpha \in [0, 1]$ let the MV-algebra S_α be defined by

$$S_\alpha = \{m + n\alpha \mid m, n \in \mathbf{Z}, 0 \leq m + n\alpha \leq 1\}.$$

Then it is not hard to see that, for any irrational $0 < \beta < 1$, $S_\alpha = S_\beta$ if and only if $\alpha = \beta$ or $\alpha = 1 - \beta$.

As a matter of fact, assume $S_\alpha = S_\beta$ and write $0 < \alpha < \beta < 1$ without loss of generality. By definition, $\alpha = m + n\beta$ and $\beta = p + q\alpha$. It follows that $\alpha = m + np + nq\alpha$, whence $m + np = 0$ and $nq = 1$. In case $n = q = 1$, we get the contradiction $0 < \beta - \alpha = p < 1$. In the remaining case when $n = q = -1$, we get $\alpha = m - \beta < 1$, whence $m = 1$, as required.

By Corollary 7.2.6, we then have

Corollary 7.2.7 *There are uncountably many nonisomorphic simple subalgebras of $[0, 1]$ with one generator.* □

7.3 The radical

Let G be an ℓ-group. Given elements a, b in G^+, we say that a is *infinitely smaller* than b, in symbols, $a \ll b$, iff $na \leq b$ for each integer $n \geq 0$.

Lemma 7.3.1 *For any ℓ-group G with strong unit u, we have:*

(7.11) $Rad(\mathbf{\Gamma}(G,u)) = \{x \in G^+ \mid x \ll u\}$.

Proof: By Proposition 3.6.4, for every MV-algebra A we have $Rad(A) = Infinit(A) \cup \{0\}$. Let $H = \{x \in G^+ \mid x \ll u\}$. If $x \in H$ then a fortiori, $x \in [0,u]$. For each integer $0 \leq n$ we have $(n+1)|x| \leq u$; using the notational convention of (2.7), we can write

$$n.|x| = u \wedge n|x| \leq u \wedge (u - |x|) = \neg x,$$

whence $H \subseteq Infinit(\mathbf{\Gamma}(G,u))$.

Conversely, let $x \in \mathbf{\Gamma}(G,u)$ be infinitesimal. By induction on n we shall prove that $nx \leq u$. The case $n = 0$ is trivial. For the induction step, suppose $nx \leq u$. Then $nx = u \wedge nx = n.x \leq u - x$, whence $(n+1)x \leq u$. In conclusion, $x \ll u$, as required to complete the proof. □

Recall from Theorem 7.2.2 the definition of the map ϕ.

Lemma 7.3.2 *Let J be an ideal of the MV-algebra $\mathbf{\Gamma}(G,u)$ such that $J \subseteq Rad(\mathbf{\Gamma}(G,u))$. Then $\phi(J) = \{x \in G \mid |x| \in J\}$.*

Proof: In the light of Proposition 3.6.4, we must prove that whenever $|x| \wedge u$ is infinitesimal then $|x| \leq u$. As a matter of fact, by Lemma 7.3.1, if $|x| \wedge u$ is infinitesimal, then $|x| \wedge u \ll u$. Since u is a strong unit of G, there is $0 \leq m \in \mathbf{Z}$ such that $|x| \leq mu$. Therefore, $|x| \leq m(|x| \wedge u) \leq u$. □

Let A be an MV-algebra and J an ideal of A contained in $Rad(A)$. With reference to Theorem 2.4.5, since $\varphi_A: A \to \mathbf{\Gamma}(G_A, u_A)$ is an isomorphism, it follows that $\varphi_A(J)$ is an ideal of $\mathbf{\Gamma}(G_A, u_A)$ contained in $Rad(\mathbf{\Gamma}(G_A, u_A))$. As an ℓ-ideal of G_A, $\phi(\varphi_A(J))$ is an ℓ-group (more precisely, an ℓ-subgroup of G_A). Since by Lemma 7.3.2, $\phi(\varphi_A(J)) \subseteq \{x \in G_A \mid |x| \ll u_A\} \subseteq [0, u_A]$, we conclude that $\varphi_A(J) = (\phi(\varphi_A(J)))^+$.

As a consequence we have

Lemma 7.3.3 *Let A be an MV-algebra and J an ideal of A such that $J \subseteq Rad(A)$. Then we have*

(i) *For all $x, y \in J$, $\varphi_A(x) \oplus \varphi_A(y) = \varphi_A(x) + \varphi_A(y)$;*

(ii) *J is* nilpotent, *i.e.*, $J^2 = \{x \odot y \mid x, y \in J\} = \{0\}$.

Proof: Suppose $x, y \in J$. Since $\varphi_A(x)$ and $\varphi_A(y)$ both belong to $(\phi(\varphi_A(J)))^+$, then $\varphi_A(x) + \varphi_A(y) \in (\phi(\varphi_A(J)))^+$. Therefore, $\varphi_A(x) + \varphi_A(y) < u_A$, whence (i) follows. Moreover, $\varphi_A(x) \odot \varphi_A(y) = (\varphi_A(x) + \varphi_A(y) - u_A) \vee 0 = 0$. Since φ_A is an isomorphism, also (ii) is settled. □

Theorem 7.3.4 *Let A be an MV-algebra. For each ideal $J \subseteq Rad(A)$ there is an ℓ-group $\mathbf{G}(J)$, unique up to isomorphism, such that $\langle J, \oplus, 0\rangle$ is isomorphic to the ordered monoid $\mathbf{G}(J)^+$ of positive elements of $\mathbf{G}(J)$.*

Proof: Put $\mathbf{G}(J) =_{def} \phi(\varphi_A(J))$. From Lemmas 7.3.2 and 7.3.3(i) it follows that φ_A is both a monoid-isomorphism and an order-isomorphism from J onto $\mathbf{G}(J)^+$. Uniqueness follows from the fact that any ℓ-group is determined by the ordered monoid of its positive elements. □

Remark: Suppose that J is a nilpotent ideal of an MV-algebra A. If $x \in J$, then $nx \in J$, and since J is nilpotent, $nx \odot nx = 0$. This implies that $nx \leq \neg nx \leq \neg x$. Therefore, $J \subseteq Rad(A)$. By Lemma 7.3.3(ii), *$J \subseteq Rad(A)$ iff J is nilpotent.* In other words, *$Rad(A)$ is the largest nilpotent ideal of A.*

7.4 Perfect MV-algebras

Let A be an MV-algebra. As an intersection of (maximal) ideals of A, $Rad(A)$ is itself an ideal of A. Let the ℓ-group $\mathbf{G}(Rad(A))$ be as in Theorem 7.3.4. If A is semisimple, then $\mathbf{G}(Rad(A)) = \{0\}$. At the other extreme, one can naturally investigate MV-algebras having as many infinitesimals as possible. This leads to the following

Definition 7.4.1 An MV-algebra A is called *perfect* iff A is nontrivial and $A = Rad(A) \cup \neg Rad(A)$, where $\neg Rad(A) = \{x \in A \mid \neg x \in Rad(A)\}$. For each perfect MV-algebra A, let

$$\mathbf{\Delta}(A) =_{def} \mathbf{G}(Rad(A)),$$

and for each homomorphism h from A into a perfect MV-algebra B, let

$$\mathbf{\Delta}(h) =_{def} \Xi(h)|_{\{x \in G_A \mid |x| \ll u_A\}}$$

be the restriction of $\Xi(h)$ to the set $\{x \in G_A \mid |x| \ll u_A\}$.

Since ℓ-group homomorphisms preserve the $\ll$ relation, it follows that $\mathbf{\Delta}(h)$ is an ℓ-group homomorphism from $\mathbf{\Delta}(A)$ into $\mathbf{\Delta}(B)$. It is easy to check that $\mathbf{\Delta}$ is a functor from the the category of perfect MV-algebras (considered as a full subcategory of the category of MV-algebras) into the category of ℓ-groups. Conversely, we shall give a method for constructing perfect MV-algebras from ℓ-groups.

To this purpose, let H be a totally ordered abelian group and G be an ℓ-group. The *lexicographic product* $H \circ G$ of H and G is the ℓ-group whose elements are the pairs (x, a) such that $x \in H$ and $a \in G$, with the group operations defined pointwise and the order relation given by $(x, a) < (y, b)$ iff either $x < y$ or $x = y$ and $a < b$. Note that the infimum of two elements in $H \circ G$ is obtained as follows:

$$(7.12)\quad (x,a) \wedge (y,b) = \begin{cases} (x,a) & \text{if} \quad x < y \\ (y,b) & \text{if} \quad y < x \\ (x, a \wedge b) & \text{if} \quad x = y \end{cases}$$

For each ℓ-group G, we write $\mathbf{\Lambda}(G) = \mathbf{Z} \circ G$, where, as usual, $\mathbf{Z}$ denotes the additive group of integers with natural order. Further, for any ℓ-group H and ℓ-group homomorphism $h: G \to H$, let us define the function $\mathbf{\Lambda}(h): \mathbf{\Lambda}(G) \to \mathbf{\Lambda}(H)$ by

$$(7.13)\quad \mathbf{\Lambda}(h)((m,a)) = (m, h(a)) \quad \text{for each pair} \quad (m,a) \in \mathbf{Z} \circ G.$$

Then $\mathbf{\Lambda}(h)$ is an ℓ-group homomorphism. As a matter of fact, $\mathbf{\Lambda}$ is a functor from the category of lattice-ordered abelian groups into itself.

Let G be an ℓ-group. Since for each $m > 0$ and each $a \in G$, the pair (m, a) is a strong unit of $\mathbf{Z} \circ G$, we can write

$$\Sigma(G) = \Gamma(\mathbf{\Lambda}(G), (1,0)).$$

The zero element of $\Sigma(G)$ is $(0,0)$ and the operations $\oplus$ and $\neg$ are defined as follows:

$$(i,a) \oplus (j,b) = \begin{cases} (0, a+b), & \text{if } \ i+j=0 \\ (1, (a+b) \wedge 0), & \text{if } \ i+j=1 \\ (1,0), & \text{if } \ i+j=2 \end{cases}$$

and

$$\neg(i,a) = (1-i, -a).$$

By definition of lexicographic order it follows that

$$\Sigma(G) = \{(0,a) \mid a \in G^+\} \cup \{(1,b) \mid b \in G^-\}.$$

The MV-algebra $\mathbf{C} = \Sigma(\mathbf{Z})$ is a first example of a nonsemisimple MV-algebra. More generally, since $Rad(\Sigma(G)) = \{(0,a) \mid a \in G\}$, and for each $a \in G$, $(1,-a) = (1,0) - (0,a) = \neg(0,a)$, the algebras $\Sigma(G)$ are examples of perfect MV-algebras.

Lemma 7.4.2 *For each perfect MV-algebra A, the stipulation*

$$\alpha_A(x) = \begin{cases} (0, \varphi_A(x)), & \textit{if } \ x \in Rad(A) \\ (1, -\varphi_A(\neg x)), & \textit{if } \ x \in \neg Rad(A) \end{cases}$$

defines an isomorphism α_A from A onto $\Sigma(\mathbf{\Delta}(A))$.

Proof: Since $\varphi_A(Rad(A)) = (\mathbf{\Delta}(A))^+$, α_A is a one-one mapping from A into $\Sigma(\mathbf{\Delta}(A))$. To see that α_A maps A onto $\Sigma(\mathbf{\Delta}(A))$, take $(i,z) \in \Sigma(\mathbf{\Delta}(A))$. If $i = 0$, then $z = \varphi_A(x)$, for some $x \in Rad(A)$, and $(0,z) = \alpha_A(x)$. If $i = 1$, then $z = -\varphi_A(y)$, with $y = \neg\neg y \in Rad(A)$. Hence $(1,z) = \alpha_A(\neg y)$. We shall now show that α_A is a homomorphism. Trivially, $\alpha_A(0) = (0,0)$, and a simple computation shows that $\alpha_A(\neg x) = (1,0) - \alpha_A(x) = \neg\alpha_A(x)$. There remains to be proved

(7.14) $\alpha_A(x \oplus y) = \alpha_A(x) \oplus \alpha_A(y)$.

Case 1: $x \oplus y \in Rad(A)$.

Then both x and y are in $Rad(A)$, and equation (7.14) follows at once from Lemma 7.3.3(i).

Case 2: $x \oplus y \in \neg Rad(A)$.

Subcase 2a: $x \in \neg Rad(A)$ and $y \in \neg Rad(A)$.

Then by Lemma 7.3.3(ii), $\alpha_A(x \oplus y) = (1, -\varphi_A(\neg x \odot \neg y)) = (1,0)$, and by definition of lexicographic order, $\alpha_A(x) \oplus \alpha_A(y) = (2, -(\varphi_A(x) + \varphi_A(y))) \wedge (1,0) = (1,0)$.

Subcase 2b: $x \in Rad(A)$ and $y \in \neg Rad(A)$.

From $\varphi_A(\neg(x \oplus y)) = (u - \varphi_A(x) + \varphi_A(\neg y) - u) \vee 0 = (\varphi_A(\neg y) - \varphi_A(x)) \vee 0$, we get $\alpha_A(x \oplus y) = (1, (\varphi_A(x) - \varphi(\neg y)) \wedge 0) = ((0, \varphi_A(x)) + (1, -\varphi_A(\neg y))) \wedge (1,0) = \alpha_A(x) \oplus \alpha_A(y)$.

Subcase 2c: $x \in \neg Rad(A)$ and $y \in Rad(A)$.

One argues as in the previous case, by interchanging the roles of x and y. □

If h is an ℓ-group homomorphism from G into an ℓ-group H, it follows that $\mathbf{\Lambda}(h): \mathbf{\Lambda}(G) \to \mathbf{\Lambda}(H)$ is an ℓ-group homomorphism such that $\mathbf{\Lambda}(h)((1,0)) = (1,0)$. Therefore the map $\mathbf{\Sigma}(h)$ defined by $\mathbf{\Sigma}(h) =_{def} \mathbf{\Gamma}(\mathbf{\Lambda}(h)): \mathbf{\Sigma}(G) \to \mathbf{\Sigma}(H)$ is a homomorphism. It is now easy to see that $\mathbf{\Sigma}$ is a functor from the category of ℓ-groups to the category of MV-algebras.

For each ℓ-group G, we have the identity $Rad(\mathbf{\Sigma}(G)) = \{0\} \times G^+$. It follows that

$$\phi((\mathbf{\Sigma}(G))) = \{(m,x) \in \mathbf{\Lambda}(G) \mid |(m,x)| \ll (1,0)\} = \{0\} \times G,$$

whence the mapping $a \mapsto \beta_G(a) = (0,a)$ is an ℓ-group isomorphism from G onto $\mathbf{\Delta}(\mathbf{\Sigma}(G))$.

The next lemma shows that the composite functor $\mathbf{\Delta\Sigma}$ is naturally equivalent to the identity functor of the category of ℓ-groups.

Lemma 7.4.3 *For all ℓ-groups G, H and for any ℓ-group homomorphism $h: G \to H$, we have a commutative diagram*

$$\begin{array}{ccc} G & \xrightarrow{h} & H \\ \beta_G \downarrow & & \downarrow \beta_H \\ \mathbf{\Delta}(\mathbf{\Sigma}(G)) & \xrightarrow{\mathbf{\Delta}(\mathbf{\Sigma}(h))} & \mathbf{\Delta}(\mathbf{\Sigma}(H)) \end{array}$$

in the sense that, for each $a \in G$, $\beta_H(h(a)) = \Delta(\Sigma(h))(\beta_A(a))$.

Proof: It is sufficient to observe that $\beta_H(h(a)) = (0, h(a))$, and

$$\Delta(\Sigma(h))(\beta_G(a))$$

$$= \Delta(\Sigma(h))(0, a) = (0, h(a)).$$

□

Conversely, the following lemma shows that the composite functor $\Sigma\Delta$ is naturally equivalent to the identity functor of the category of MV-algebras:

Lemma 7.4.4 *For all MV-algebras A, B and for any homomorphism* $h: A \to B$, *we have a commutative diagram*

$$\begin{array}{ccc} A & \xrightarrow{h} & B \\ \alpha_A \downarrow & & \downarrow \alpha_B \\ \Sigma(\Delta(A)) & \xrightarrow{\Sigma(\Delta(h))} & \Sigma(\Delta(B)) \end{array}$$

in the sense that, for each $a \in A$, $\alpha_B(h(a)) = \Sigma(\Delta(h))(\alpha_A(a))$.

Proof: If $a \in Rad(A)$ then, trivially, $h(a) \in Rad(B)$ and $\alpha_B(h(a)) = (0, \varphi_B(h(a)))$. On the other hand,

$$\Sigma(\Delta(h))(\alpha_A(a)) = \Sigma(\Delta(h)(0, \varphi_A(a))) = (0, \Gamma(\Xi(h))(\varphi_A(a))),$$

and the result follows from Theorem 7.1.2. □

From Lemmas 7.4.3 and 7.4.4 we immediately get:

Theorem 7.4.5 *The* Δ *functor establishes a natural equivalence from the category of perfect MV-algebras, considered as a full subcategory of the category of MV-algebras, and the category of* ℓ*-groups.* □

7.5 Bibliographical remarks

The natural equivalence between MV-algebras and ℓ-groups with strong unit was first established in [163], building on previous work by Chang [38] for the totally ordered case (see also [131]).

The proof of Theorem 7.2.4 can be found in [56]. The results established in the remark following Lemma 7.3.3 are due to Belluce [14].

As a generalization of the MV-algebra algebra **C** originally introduced by Chang in [36], perfect MV-algebras were introduced in [20]. The natural equivalence between the categories of perfect MV-algebras and ℓ-groups was established by Di Nola and Lettieri [65] (see also [60]).

Chapter 8

Varieties of MV-algebras

A class $\mathcal{C}$ of MV-algebras is said to be a *variety* (or, an *equational class*), iff there is a set $\mathcal{E}$ of MV-equations such that for every MV-algebra A, $A \in \mathcal{C}$ iff A satisfies all equations in $\mathcal{E}$. For instance, when $\mathcal{E} = \emptyset$, we obtain the variety $\mathcal{MV}$ of MV-algebras. When $\mathcal{E} = \{x = y\}$, we obtain the variety of trivial MV-algebras. The main aim of this chapter is to describe all varieties of MV-algebras.

8.1 Basic definitions

A variety $\mathcal{C}$ of MV-algebras is called *proper* iff it is nontrivial and is different from $\mathcal{MV}$. By Corollary 1.5.5, the class of boolean algebras is the proper equational class of MV-algebras determined by $\mathcal{E} = \{x = x \oplus x\}$. By Lemma 1.4.5 every variety $\mathcal{C}$ is closed under subalgebras, homomorphic images and direct products. Using this observation we obtain that the following three classes of MV algebras are not varieties: semisimple, hyperarchimedean, and complete MV algebras. Indeed, since by Corollary 3.6.2 all free MV-algebras are semisimple, from Proposition 3.1.5 it follows that the class of semisimple MV-algebras is not closed under quotients. As noted after Definition 6.3.1, the class of hyperarchimedean MV algebras is not closed under direct products. Finally, the class of complete MV algebras is not closed under subalgebras, because $\mathbf{Q} \cap [0, 1]$ is an incomplete subalgebra of the complete MV algebra [0,1].

Let $\mathcal{K}$ be a set of MV-algebras and $\mathcal{E}$ be the set of all MV-equations which are satisfied by all MV-algebras in $\mathcal{K}$. The variety determined by $\mathcal{E}$ is said to be *generated by* $\mathcal{K}$, and is denoted by $\mathcal{C}(\mathcal{K})$. When the set $\mathcal{K}$ consists of one MV-algebra A, we shall write $\mathcal{C}(A)$ instead of $\mathcal{C}(\{A\})$, and we will call $\mathcal{C}(A)$ *the variety generated by* A. With this notation, the Completeness Theorem 2.5.3 amounts to the identity $\mathcal{MV} = \mathcal{C}([0,1])$. More generally we have

Proposition 8.1.1 *If A is an infinite subalgebra of the MV-algebra $[0,1]$, then $\mathcal{C}(A) = \mathcal{MV}$.*

Proof: Since the operations $\neg x = 1 - x$ and $x \oplus y = \min(1, x + y)$ are continuous over $[0,1]$, then for each MV-term $\tau(x_1, \ldots, x_n)$, the function $\tau^{[0,1]}\colon [0,1]^n \to [0,1]$ is continuous. Therefore, since by Proposition 3.5.3, A is dense in $[0,1]$, the equation $\tau(x_1, \ldots, x_n) = 0$ holds in $[0,1]$ iff it holds in A. The desired result is now a consequence of the Completeness Theorem. □.

Proposition 8.1.2 *If $2 \leq n_0 < n_1 < \ldots$ is an infinite sequence of natural numbers, then $\mathcal{C}(\{\text{Ł}_{n_i} \mid i = 0, 1, \ldots\}) = \mathcal{MV}$.*

Proof: For each $a \in [0,1]$ there is a number k such that $1 \leq k \leq n - 1$ and $|a - \frac{k}{n-1}| < \frac{1}{n-1}$. The desired result now follows from the continuity of the term functions $\tau^{[0,1]}\colon [0,1]^n \to [0,1]$. □

As a consequence of Theorems 1.3.3 and 1.4.6, *every variety of MV-algebras is generated by the collection of its MV-chains.*

If A is a proper MV-chain, $A/\mathit{Rad}(A)$ is a simple MV-algebra, because $\mathit{Rad}(A)$ is the unique maximal ideal of A. By Theorem 3.5.1, $A/\mathit{Rad}(A)$ is isomorphic to a subalgebra of the MV-algebra $[0,1]$.

Definition 8.1.3 Let A be a proper MV-chain. If for some $n \geq 2$, $A/\mathit{Rad}(A) \cong \text{Ł}_n$, then we say that A is of *rank* n. If $A/\mathit{Rad}(A)$ is isomorphic to an infinite subalgebra of $[0,1]$, we say that A is of *infinite rank.*

As an example, for each integer $n \geq 2$ the MV-algebra Ł_n is a simple MV-chain of rank n.

For every totally ordered abelian group G, the algebra $\Sigma(G) = \Gamma(\Lambda(G),(1,0))$ introduced in Section 7.4 is a nonsimple MV-chain of rank 2. More generally, for each $b \in G$ and each integer $n \geq 2$, consider the MV-algebra $\Gamma(\Lambda(G),(n-1,b))$. By definition of lexicographic product, $\Gamma(\Lambda(G),(n-1,b))$ coincides with the union $A \cup B \cup C$, where A is the set of all pairs of the form $(0,x)$, for arbitrary $b \leq x \in G$, B is the set of all pairs $(n-1,y)$, for arbitrary $b \geq y \in G$, and $C = \bigcup_{i=1}^{n-2}(\{i\} \times G)$. The pair $(0,0)$ is the zero element, and the operations $\neg$ and $\oplus$ given by

$$(8.1)\quad (i,x) \oplus (j,y) = \begin{cases} (i+j, x+y) & \text{if } i+j < n-1 \text{ or if} \\ & i+j = n-1 \text{ and } x+y < b \\ (n-1,b) & \text{otherwise.} \end{cases}$$

$$(8.2)\quad \neg(i,x) = (n-1-i, b-x)$$

It follows that the $\odot$ operation of $\Gamma(\Lambda(G),(n-1,b))$ is given by

$$(8.3)\quad (i,x) \odot (j,y) = \begin{cases} (i+j-n+1, x+y-b) & \text{if } i+j > n-1 \\ & \text{or if } i+j = n-1 \\ & \text{and } x+y > b. \\ (0,0) & \text{otherwise.} \end{cases}$$

A straightforward verification shows that $\Gamma(\Lambda(G),(n-1,b))$ is a nonsimple MV-chain of rank n. Indeed, the projection $\pi_1 \colon \Lambda(G) \to \mathbf{Z}$, defined by $\pi_1((j,t)) = j$, for each $(j,t) \in \Lambda(G)$, is a surjective ℓ-group homomorphism, with $\pi_1((n-1,b)) = n-1$. Hence

$$\Gamma(\pi_1) \colon \Gamma(\Lambda(G),(n-1,b)) \to \Gamma(\mathbf{Z}, n-1)$$

is a surjective homomorphism. Since

$$Ker(\Gamma(\pi_1)) = \{(0,t) \mid t \in G^+\} = Rad(\Gamma(\Lambda(G),(n-1,b))),$$

we have

$$\Gamma(\Lambda(G),(n-1,b))/Rad(\Gamma(\Lambda(G),(n-1,b))) \cong \Gamma(\mathbf{Z}, n-1) \cong \text{Ł}_n.$$

For each integer $n \geq 2$, we shall write, in abbreviated form,

$$\mathrm{K}_n = \mathbf{\Gamma}(\mathbf{\Lambda}(\mathbf{Z}), (n-1, 0))$$

and

$$\mathrm{H}_n = \mathbf{\Gamma}(\mathbf{\Lambda}(\mathbf{Z}), (n-1, 1)).$$

Proposition 8.1.4 *If C is a proper variety of MV-algebras then there is an integer $m \geq 2$ such that $rank(A) \leq m$ for all MV-chains $A \in C$.*

Proof: Immediate from Propositions 8.1.1 and 8.1.2, upon noting that $A \in C$ implies $A/Rad(A) \in C$. □

In the next sections we shall establish that, conversely, if the ranks of all MV-chains in a variety C are bounded, then $C \neq \mathcal{MV}$. We first consider simple MV-algebras of finite rank, i.e., the algebras Ł_n. The simplicity assumption shall be removed in a subsequent section.

8.2 Varieties from simple algebras

Following the notation in Section 1 of Chapter 2 (see Equation (2.7)), for $x \in [0, 1]$, we will write $n.x$ instead of $x \oplus \cdots \oplus x$ (n times), and write nx instead of $x + \cdots + x$ (n times).

Definition 8.2.1 Let $n \geq 2$ be an integer. By an *n-bounded* MV-algebra we shall mean an algebra satisfying the equation

(E_{n0}) $\quad (n-1).x = n.x$

The variety of n-bounded MV-algebras will be denoted by $\mathcal{U}_n$.

Theorem 8.2.2 *Let A be an MV-algebra and $n \geq 2$ an integer. Then $A \in \mathcal{U}_n$ if and only if A is a subdirect product of algebras Ł_k, with $2 \leq k \leq n$.*

Proof: Let $A \in \mathcal{U}_n$. By the remark following Definition 3.6.3 we have $Rad(A) = (0)$. Hence A is semisimple, i.e., A is a subdirect product of subalgebras of $[0, 1]$. By Theorem 1.4.6, all these subalgebras satisfy the equation $(n-1).x = n.x$. Let S be a subalgebra of [0,1] such that, for each $k = 2, 3, \ldots, n$, $S \neq Ł_k$. Then by Proposition 3.5.3 there is $t \in S$ such that $0 < t < 1/(n-1)$. It follows that $(n-1).t = \min(1, (n-1)t) < \min(1, nt) = n.t$. Therefore A is a subdirect product of algebras $\mathbf{Ł}_k$, for suitable integers $2 \leq k \leq n$.

Conversely, let A be a subdirect product of algebras $\mathbf{Ł}_k$, with $2 \leq k \leq n$. For all $x \in \mathbf{Ł}_k$, we either have $x = 0$ or $x \geq 1/(k-1)$. From $k-1 \leq n-1$ it follows that min $(1, (n-1)x) = 1 =$ min $(1, nx)$. Thus, for each $2 \leq k \leq n$ the MV-chain $\mathbf{Ł}_k$ belongs to $\mathcal{U}_n$; by Theorem 1.4.6, $A \in \mathcal{U}_n$. □

Recalling now Theorem 1.4.6 we have

Corollary 8.2.3 *For each integer $n \geq 2$, $\mathcal{U}_n = \mathcal{C}(\{\mathbf{Ł}_2, \mathbf{Ł}_3, \ldots, \mathbf{Ł}_n\})$.* □

Corollary 8.2.4 *An MV-algebra A belongs to the variety $\mathcal{C}(\mathbf{Ł}_2)$ iff A is a boolean algebra, i.e., iff A satisfies the equation $x \oplus x = x$. An MV-algebra belongs to $\mathcal{C}(\mathbf{Ł}_3)$ iff it satisfies the equation $x \oplus x = x \oplus x \oplus x$.*

Proof: The first result easily follows from Theorems 8.2.2 and 1.5.3(iv). For the second result, by the remark following Proposition 3.5.3, $\mathbf{Ł}_2$ is a subalgebra of $\mathbf{Ł}_3$. Thus, an equation is simultaneously satisfied by $\mathbf{Ł}_2$ and by $\mathbf{Ł}_3$ iff it is satisfied by $\mathbf{Ł}_3$. Thus, $\mathcal{C}(\{\mathbf{Ł}_2, \mathbf{Ł}_3\}) = \mathcal{C}(\mathbf{Ł}_3)$, and the desired conclusion now follows from Theorem 8.2.2. □

8.3 MV-chains of finite rank

As we have seen above, for every totally ordered group G, algebras of the form $\Gamma(\mathbf{\Lambda}(G), (n-1, b))$ yield examples of MV-chains of rank n. We shall see in this section that these are in fact the most general examples of nonsimple MV-chains of rank n. Firstly we introduce some notations.

Let A be a totally ordered MV-algebra. By Theorem 7.3.4 we can associate with $Rad(A)$ the ℓ-group $\mathbf{G}(A)$, which is an ℓ-subgroup of the

ℓ-group G_A. Hence, by the remark following Theorem 2.4.5, $\mathbf{G}(Rad(A))$ is a totally ordered group, denoted $\rho(A)$. When A is a perfect MV-chain, then $\rho(A) = \mathbf{\Delta}(A)$.

Theorem 8.3.1 *Let A be a nonsimple MV-chain of rank n. Then there is $0 < b \in Rad(A)$ such that $A \cong \mathbf{\Gamma}(\mathbf{\Lambda}(\rho(A)), (n-1, \varphi_A(b)))$.*

Proof: Let $f: A \to \mathbf{L}_n$ be the composition of the natural homomorphism from A onto $A/Rad(A)$ and the isomorphism from $A/Rad(A)$ onto $\mathbf{L}_n$. By Lemma 7.2.1, there is a unique ℓ-group homomorphism $g: G_A \to \mathbf{Z}$ such that $g(u_A) = n-1$ and $\mathbf{\Gamma}(g) = f$. Since f is surjective, by the same Lemma we know that g is surjective. Since f is not injective (because A is not simple), g is not injective, that is, $Ker(g) = \{x \in G_A \mid g(x) = 0\} \neq \{0\}$. By Theorem 7.2.4, we have isomorphisms

$$\mathbf{\Gamma}(G_A, u_A)/\varphi_A(Rad(A))$$

$$\cong \mathbf{\Gamma}(\mathbf{Z}, (n-1))$$

$$\cong \mathbf{\Gamma}(G_A/Ker(g), u_A/Ker(g))$$

$$\cong \mathbf{\Gamma}(G_A, u_A)/\psi(Ker(g)).$$

Since every (nontrivial) MV-chain has exactly one maximal ideal, we obtain $\rho(A) = \phi(\varphi_A(Rad(A))) = \phi(\psi(Ker(g))) = Ker(g)$. Therefore, to prove the theorem it is sufficient to prove the following

Claim. There is an ℓ-group isomorphism $h: G_A \to \mathbf{\Lambda}(Ker(g))$ such that $h(u_A) = (n-1, v)$, for some $0 < v \in Ker(g)$.

To prove the claim, we first show that there is $0 < z \in G_A$ such that $g(z) = 1$ and $(n-1)z < u_A$. Indeed, since g is surjective, there is $y \in G_A$ such that $g(y) = 1$. Note that $y > 0$, for otherwise $g(y) \leq g(0) = 0$, which is impossible. If $(n-1)y < u_A$, then we can put $z = y$. If $(n-1)y = u_A$, then we can put $z = y - t$, where $t \in Ker(g)$, $t > 0$ (note that $t < y$, for otherwise $g(y) \leq g(t) = 0$, a contradiction). Finally, if $(n-1)y > u_A$, then $(n-1)y = u + s$, for some $0 < s \in Ker(g)$, and we can put $z = y - 2s$. To complete the proof of the claim it is enough to let h be defined by $h(x) = (g(x), x - g(x)z)$, for each $x \in G_A$, and $v = u_A - (n-1)z$. □

Lemma 8.3.2 *For each integer $n \geq 2$ and each nonsimple MV-chain A of order n, $\mathrm{H}_n \in \mathcal{C}(A)$.*

Proof: By Lemma 8.3.1 we can assume $A \cong \Gamma(\Lambda(G), b)$ for some totally ordered group G and $0 < b \in G$. The mapping $h: \Lambda(\mathbf{Z}) \to \Lambda(G)$ defined by $h((j,m)) = (j, mb)$, for each $(j,m) \in \mathbf{Z} \times \mathbf{Z}$, is an injective ℓ-group homomorphism and $h((n-1,1)) = (n-1,b)$. Therefore $\Gamma: \mathrm{H}_n \to \Gamma(\Lambda(G),(n-1,b))$ is an injective homomorphism, whence it follows that $\mathrm{H}_n \in \mathcal{C}(A)$. □

For all integers k and n, $n \geq 2$, let the map $h: \Lambda(\mathbf{Z}) \to \Lambda(\mathbf{Z})$ be defined by $h((j,m)) = (j,(n-1)m - kj)$. Then h is an injective ℓ-group homomorphism with $h((n-1,k)) = (n-1,0)$. It follows that

$$\Gamma(h): \Gamma(\Lambda(\mathbf{Z}),(n-1,k)) \to \mathrm{K}_n$$

is an embedding, whence

(8.4) for all integers k and n, $n \geq 2$, $\Gamma(\Lambda(\mathbf{Z}),(n-1,k)) \in \mathcal{C}(\mathrm{K}_n)$.

This result is considerably strengthened by the following proposition:

Proposition 8.3.3 *Let G be an ℓ-group and $0 < b \in G$. Then for each integer $n \geq 2$, $\Gamma(\Lambda(G),(n-1,b)) \in \mathrm{K}_n$.*

Proof: We need the following classical result from the theory of ℓ-groups:

For any lattice-ordered abelian group G there exists a nonempty set I, an ℓ-subgroup H of the ℓ-group $\mathbf{Z}^I$ and an ℓ-group homomorphism from H onto G.

We sketch here a proof for the reader's convenience. Since G is a subdirect product of totally ordered abelian groups (see the proof of the Claim in Lemma 7.1.5), one can easily adapt the proof of the Completeness Theorem (Theorem 2.5.3) to show that an ℓ-group equation holds in all ℓ-groups iff it holds in $\mathbf{Z}$. For each nonempty set I let H be the ℓ-subgroup of the ℓ-group $\mathbf{Z}^I$ generated by the set P of projection functions $\pi_i: \mathbf{Z}^I \to \mathbf{Z}$, for all $i \in I$. Arguing as in the proofs of Lemma 3.1.2 and Proposition 3.1.4, one immediately shows that H has the following property: *Each function from the set P to an ℓ-group G can be uniquely extended to an ℓ-group homomorphism from H to G.* The desired conclusion now follows by letting I be an arbitrary set whose cardinality is greater or equal than that of G (compare with Proposition 3.1.5).

Returning to the proof of the proposition, suppose that G is an ℓ-group and $0 < b \in G$. Let H an ℓ-subgroup of $\mathbf{Z}^I$ and $h: H \to G$ be a surjective ℓ-group homomorphism. Then there is $c \in H$ such that $h(c) = b$; replacing, if necessary, c by $c \vee 0$, we can safely assume $c > 0$. It follows that $\mathbf{\Lambda}(h): \mathbf{\Lambda}(H) \to \mathbf{\Lambda}(G)$ is a surjective ℓ-group homomorphism and $\mathbf{\Lambda}(h)((n-1, c)) = (n-1, b)$. Hence

$$\mathbf{\Gamma}(h): \mathbf{\Gamma}(\mathbf{\Lambda}(H), (n-1, c)) \to \mathbf{\Gamma}(\mathbf{\Lambda}(G), (n-1, b))$$

is a surjective homomorphism. Since H is an ℓ-subgroup of $\mathbf{Z}^I$, each element of $x \in H$ is a function $x: I \to \mathbf{Z}$. The function $f: \mathbf{\Lambda}(H) \to \mathbf{\Lambda}(\mathbf{Z})^I$ defined by $f((m, x))(i) = (m, x(i))$, for $i \in I$, $x \in H$ and $m \in \mathbf{Z}$, is an injective ℓ-group homomorphism. Let $J = \{i \in I \mid c(i) \neq 0\}$. Since $c > 0$, J is nonempty; moreover, $i \in J$ iff $c(i) > 0$. Therefore $\mathbf{\Gamma}(\mathbf{\Lambda}(H), (n-1, c))$ is isomorphic to a subalgebra of the product algebra

$$\Pi_{i \in J} \mathbf{\Gamma}(\mathbf{\Lambda}(\mathbf{Z}), (n-1, c(i))).$$

By (8.4), for each $i \in J$, we get $\mathbf{\Gamma}(\mathbf{\Lambda}(\mathbf{Z}), (n-1, c(i))) \in \mathcal{C}(\mathrm{K}_n)$. Therefore,

$$\mathbf{\Gamma}(\mathbf{\Lambda}(\mathbf{Z}), (n-1, c)) \in \mathcal{C}(\mathrm{K}_n).$$

Since the MV-algebra $\mathbf{\Gamma}(\mathbf{\Lambda}(G), (n-1, b))$ is a homomorphic image of $\mathbf{\Gamma}(\mathbf{\Lambda}(\mathbf{Z}), (n-1, c))$, we finally get

$$\mathbf{\Gamma}(\mathbf{\Lambda}(G), (n-1, b)) \in \mathcal{C}(\mathrm{K}_n),$$

as required to complete the proof. □

Lemma 8.3.4 *For each integer $n \geq 2$, $\mathrm{K}_n \in \mathcal{C}(\mathrm{H}_n)$.*

Proof: Suppose that an equation does not hold in K_n. By the remark following Definition 1.4.4 we can assume that the equation has the form

$\pi(x_1,\ldots,x_k) = 0$, for some MV-term π in the variables $x_1,\ldots,x_k$. There are elements $c_1,\ldots,c_k \in \mathrm{K}_n$ such that

(8.5) $\pi^{\mathrm{K}_n}(c_1,\ldots,c_k) \geq (0,1)$.

For each integer $m \geq 1$, let $f_m\colon \mathbf{\Lambda}(\mathbf{Z}) \to \mathbf{\Lambda}(\mathbf{Z})$ be the ℓ-group homomorphism defined by $f_m((v,w)) = (v,mw)$ for all pairs $(v,w) \in \mathbf{Z}\times\mathbf{Z}$. For every MV-term σ in the variables $x_1,\ldots,x_k$ let $g(\sigma)$ denote the total number of symbols $\neg$ and $\oplus$ occurring in σ. By induction on $g(\sigma)$, for each k-tuple $(b_1,\ldots,b_k) \in \mathrm{K}_n \subseteq \mathrm{H}_n$ we shall prove the following inequalities:

(8.6) $f_m(\sigma^{\mathrm{K}_n}(b_1,\ldots,b_k)) - (0,g(\sigma))$

(8.7) $\leq \sigma^{\mathrm{H}_n}(f_m(b_1),\ldots,f_m(b_k))$

(8.8) $\leq f_m(\sigma^{\mathrm{K}_n}(b_1,\ldots,b_k)) + (0,g(\sigma))$.

If $g(\sigma) = 0$, then $\sigma = x_i$, for some $i \in \{1,2,\ldots,k\}$ or $\sigma = 0$. In this latter case, the result is obvious. In the former case, if $b_i = (v,w)$ then

$$f_m(\sigma^{\mathrm{K}_n}(b_1,\ldots,b_k)) = f_m(b_i) = (v,mw) = \sigma^{\mathrm{H}_n}(f_m(b_1),\ldots,f_m(b_k)),$$

thus establishing (8.6) for $g(\sigma) = 0$.

Proceeding by induction, pick an integer $d > 0$, and suppose that (8.6) holds for all MV-terms ξ in the variables $x_1,\ldots,x_k$ such that $g(\xi) < d$.

Let $\sigma(x_1,\ldots,x_k)$ be an MV-term such that $g(\sigma) = d$. From Section 1.4 we know that precisely one of the following two cases must occur:

Case 1: There is an MV-term $\tau(x_1,\ldots,t_k)$ such that $\sigma = \neg\tau$;

Case 2: There are MV-terms $\mu(x_1,\ldots,x_k)$ and $\nu(x_1,\ldots,x_k)$ such that $\sigma = \mu \oplus \nu$.

For notational convenience, and without any essential loss of generality, we shall limit ourselves to considering MV-terms in one variable.

In Case 1, we can write $g(\tau) = g(\sigma) - 1 < d$, whence, by induction hypothesis,

$$\sigma^{\mathrm{H}_n}(f_m(b))$$

$$= (n-1,1) - \tau^{\mathrm{H}_n}(f_m(b))$$

$$\leq (n-1,1) - f_m(\tau^{\mathrm{K}_n}(b)) + (0, g(\tau))$$

$$= (n-1,0) - f_m(\tau^{\mathrm{K}_n}(b)) + (0, 1+g(\tau))$$

$$= f_m(\sigma^{\mathrm{K}_n}(b)) + (0, g(\tau))$$

and

$$\sigma^{\mathrm{H}_n}(f_m(b)) \geq (n-1,1) - f_m(\tau^{\mathrm{K}_n}(b)) - (0, g(\tau))$$

$$\geq f_m(\tau^{\mathrm{K}_n}(b)) - (0, g(\sigma)).$$

From these inequalities one immediately sees that σ satisfies (8.6).

In Case 2 we have $g(\sigma) = g(\mu) + g(\nu) + 1$. By induction hypothesis,

$$\sigma^{\mathrm{H}_n}(f_m(b)) = (n-1,1) \wedge (\mu^{\mathrm{H}_n}(f_m(b)) + \nu^{\mathrm{H}_n}(f_m(b)))$$

$$\leq (n-1,1) \wedge (f_m(\mu^{\mathrm{K}_n}(b)) + f_m(\nu^{\mathrm{K}_n}(b)) + (0, g(\mu)) + (0, g(\nu)))$$

$$\leq (n-1,\ g(\mu) + g(\nu) + 1)$$

$$\wedge (f_m(\mu^{\mathrm{K}_n}(b)) + f_m(\nu^{\mathrm{K}_n}(b)) + (0, g(\mu) + g(\nu) + 1))$$

$$= ((n-1,0) \wedge (f_m(\mu^{\mathrm{K}_n}(b)) + f_m(\nu^{\mathrm{K}_n}(b))) + (0, g(\mu) + g(\nu) + 1)$$

$$= f_m(\sigma^{\mathrm{K}_n}(b)) + (0, g(\sigma))$$

and

$$\sigma^{\mathrm{H}_n}(f_m(b))$$

$$\geq (n-1,1) \wedge (f_m(\mu^{\mathrm{K}_n}(b)) + f_m(\nu^{\mathrm{K}_n}(b)) - (0, g(\mu)) - (0, g(\nu)))$$

$$\geq (n-1, 1 - g(\mu) - g(\sigma))$$

$$\wedge (f_m(\mu^{\mathrm{K}_n}(b)) + f_m(\nu^{\mathrm{K}_n}(b)) - (0, g(\mu)) - (0, g(\nu)))$$

$$= ((n-1,1) \wedge (f_m(\mu^{\mathrm{K}_n}(b)) + f_m(\nu^{\mathrm{K}_n}(b))) - (0, g(\mu) + g(\nu))$$

$$\geq ((n-1,0) \wedge (f_m(\mu^{\mathrm{K}_n}(b)) + f_m(\nu^{\mathrm{K}_n}(b))) - (0, g(\mu) + g(\sigma) + 1)$$
$$= f_m(\sigma^{\mathrm{K}_n}(b)) - (0, g(\sigma)).$$

Hence the inequalities (8.6) also hold in this case. From (8.5) and (8.6) we obtain

$$(0, m - g(\pi)) \leq f_m(\pi^{\mathrm{K}_n}(b_1, \ldots, b_k)) - (0, g(\pi))$$
$$\leq \pi^{\mathrm{H}_n}(f_m(b_1), \ldots, f_m(b_k)).$$

If $m > g(\pi)$, then $\pi^{\mathrm{H}_n}(f_m(b_1), \ldots, f_m(b_k)) > (0,1)$ and H_n does not satisfy the equation $\pi(x_1, \ldots, x_k) = 0$. We conclude that K_n must satisfy all equations that are satisfied by H_n. □

From Theorem 8.3.1, Proposition 8.3.3 and Lemmas 8.3.2 and 8.3.4, we obtain

Theorem 8.3.5 *For each integer $n \geq 2$ and each nonsimple MV-chain A of rank n, $\mathcal{C}(A) = \mathcal{C}(\mathrm{K}_n)$.* □

8.4 Komori's classification

The two formulas (8.1) and (8.3) show that, for each $(j, a) \in \mathrm{K}_n = \Gamma(\Lambda(\mathbf{Z}), (n-1, 0))$ and integer $k \geq 1$,

$$(8.9)\quad k.(j,a) = \begin{cases} (kj, ka) & \text{if } kj < n-1 \text{ or } kj = n-1 \text{ and} \\ & a < 0 \\ (n-1, 0) & \text{otherwise} \end{cases}$$

and

$$(8.10)\quad (j,a)^k = \begin{cases} (kj - (k-1)(n-1), ka) & \text{if } kj > (k-1)(n-1) \text{ or} \\ & kj = (k-1)(n-1) \text{ and} \\ & a > 0 \\ (0,0) & \text{otherwise.} \end{cases}$$

Lemma 8.4.1 *Let $n \geq 2$ be an integer and A an MV-chain. Then A has rank $\leq n$ if and only if A satisfies the equation*

(F_{n0}) $\qquad (n.x^{n-1})^2 = 2.x^n.$

Proof: Let $(j,a) \in \mathrm{K}_m$. Whenever $0 \le j \le m-2$, from (8.10) it follows that $(j,a)^{m-1} \in Rad(\mathrm{K}_m)$, $(j,a)^m = (0,0)$, and $(m-1,b)^{m-1} = (m-1,b)^m = (m-1,0)$. Hence the equation (F_{n0}) holds in the algebra K_m for $m = 2,\ldots,n$; further, by Theorem 8.3.5, (F_{n0}) holds in all MV-chains of rank $m \le n$. Since $A/Rad(A) \in \mathcal{C}(A)$, to complete the proof it is sufficient to show that for each $m > n$, the MV-algebra $\mathbf{Ł}_m$ does not satisfy equation (F_{n0}). To this purpose, we must prove that there is $t \in \mathbf{Ł}_m$ such that $n.t^{n-1} = 1$ and $2.t^n < 1$. Note that for each $t \in \mathbf{Ł}_m$, $n.t^{n-1} = 1$ iff $n((n-1)t-(n-2)) \ge 1$ and $2.t^n < 1$ iff $2(nt-(n-1)) < 1$. Therefore, for each $m > n$ we must find an integer k such that

$$\frac{n-1}{n} < \frac{k}{m-1} < \frac{2n-1}{2n}.$$

In case $n < m \le 2n$ we can take $k = m-2$. In case $2n < m$, there are integers q and r such that $q \ge 1$, $0 < r < 2n$ and $m = 2nq + r$. Thus, letting $k = m-(q+2)$ we obtain the desired conclusion. □

Theorem 8.4.2 *For each nontrivial MV-algebra A and integer $n \ge 2$, the following conditions are equivalent:*

(i) A satisfies the equation

(F_{n0}) $\qquad (n.x^{n-1})^2 = 2.x^n;$

(ii) $A \in \mathcal{C}(\{\mathrm{K}_2, \mathrm{K}_3, \ldots, \mathrm{K}_n\})$;

(iii) $A/Rad(A) \in \mathcal{C}(\{\mathbf{Ł}_2, \mathbf{Ł}_3, \ldots, \mathbf{Ł}_n\})$.

Proof: The equivalence between (i) and (ii) easily follows from the above Lemma, together with Theorems 1.3.3 and 1.4.6. To obtain the equivalence between (ii) and (iii) note that $A/Rad(A)$ is semisimple and that $\mathbf{Ł}_2, \ldots, \mathbf{Ł}_n$ is the complete list of all simple algebras of rank at most n. □

Corollary 8.4.3 *For each nontrivial MV-algebra A we have:*

(i) *$A \in \mathcal{C}(\mathrm{K}_2)$ iff A satisfies the equation*

$$(F_{20}) \qquad (x \oplus x)^2 = x^2 \oplus x^2;$$

(ii) *$A \in \mathcal{C}(\mathrm{K}_3)$ iff A satisfies the equation*

$$(F_{30}) \qquad (x^2 \oplus x^2 \oplus x^2)^2 = x^3 \oplus x^3.$$

Proof: (i) is an immediate consequence of Proposition 8.4.2, and (ii) follows from the same proposition and the fact that K_2 is a subalgebra of K_3. □

Remark: It follows from (i) in the above corollary that *all perfect MV-algebras are in $\mathcal{C}(\mathrm{K}_2)$.*

From Proposition 8.1.4 we know that each proper variety of MV-algebras is generated by a set of MV-chains of bounded rank. From Theorems 8.2.2 and 8.4.2 we conclude that, conversely, each set of MV-chains of bounded rank generates a proper variety. By Theorem 8.3.5, we then obtain

Theorem 8.4.4 *A class $\mathcal{C}$ of MV-algebras is a proper variety iff there are two finite sets I and J of integers ≥ 2 such that $I \cup J$ is nonempty and $\mathcal{C} = \mathcal{C}(\{\text{Ł}_i\}_{i \in I}, \{\mathrm{K}_j\}_{j \in J})$.*□

Notation: For each integer $n \geq 2$ we let

$$\mathbf{Div}(n)$$

be the set of all divisors $d \geq 1$ of n.

Lemma 8.4.5 *Let m and n be integers such that $n \geq m \geq 3$. For each integer p such that $2 \leq p < m - 1$, the MV-algebra K_m satisfies the equations:*

$$(E_{np}) \qquad (p.x^{p-1})^n = n.x^p$$

and

$$(F_{np}) \qquad n.x^p = (n+1).x^p$$

if and only if p does not divide $m-1$.

Proof: Let $y = (j,a)$ be an arbitrary element of K_m. Let us assume $p \notin \mathbf{Div}(m-1)$. If $y^{p-1} = 0$, then $y^p = 0$, whence $(p.y^{p-1})^n = n.y^p$ and $n.y^p = (n+1).y^p$. If, on the other hand, $y^{p-1} > 0$, then there are two subcases to consider:

If $y^p = 0$, from (8.10), together with our standing hypothesis that p does not divide $m-1$, it follows that $jp < (m-1)(p-1)$. Therefore, $p(p-1)j - p(p-2)(m-1) \leq m-2$. Hence from (8.9), we have

$$p.y^{p-1} = (p(p-1)j - p(p-2)(m-1), p(p-1)a)$$

$$\leq (m-2, p(p-1)a),$$

and again by (8.10), $(p.y^{p-1})^n = 0 = n.y^p$.

If $y^p > 0$ then $jp > (m-1)(p-1)$, whence $p.y^{p-1} = 1 = y^p$ and

$$n.y^p = n.(pj - (p-1)(m-1), pa) = (m-1, 0) = (n+1).y^p.$$

Therefore both equations (E_{np}) and (F_{np}) hold in K_m.

Conversely, let us assume that $n \geq p \geq 2$. If $y = \frac{p-1}{p}$, then $y^p = 0$ and $y^{p-1} = \frac{1}{p}$. Therefore $\mathbf{L}_{p+1}$ does not satisfy equation $(E_{n(p-1)})$, and since $\mathbf{L}_{p+1}$ is isomorphic to the subalgebra $\{(0,0), (1,0), \ldots, (p,0)\} \subseteq K_{p+1}$, we also obtain that equation $(E_{n(p-1)})$ does not hold in K_{p+1}. On the other hand, from $z = (p,-1) \in K_{p+1}$ and $z^p = (p,-p)$ we get $n.z^p < (n+1).z^p$, whence equation (F_{np}) does not hold in K_{p+1}. To complete the proof it is sufficient to note that if $p \in \mathbf{Div}(m-1)$ then K_{p+1} is a subalgebra of K_m. □

Theorem 8.4.6 *Let I and J be sets of integers ≥ 2 such that $I \cup J$ is nonempty; suppose that whenever $r, s \in I \cup J$ and $r < s$ then $r - 1 \notin$ $\mathbf{Div}(s-1)$. Let n be the greatest element of $I \cup J$. Let A be a nontrivial MV-algebra. Then*

$$A \in C(\{\mathbf{L}_i\}_{i\in I}, \{K_j\}_{j\in J})$$

iff A satisfies the equations

$$(F_{n0}) \qquad (n.x^{n-1})^2 = 2.x^n,$$

$$(E_{np}) \qquad (p.x^{p-1})^n = n.x^p,$$

for each integer p *such that* $2 \leq p < k$ *and* $p \notin \bigcup_{r \in I \cup J} \mathbf{Div}(r-1)$,

as well as the equations

$$(F_{nq}) \qquad k.x^q = (k+1).x^q,$$

for each integer $q \geq 2$ *such that*

$$q \in \bigcup_{r \in I} (\mathbf{Div}(r-1) \setminus \bigcup_{s \in J} \mathbf{Div}(s-1)).$$

Proof: Let C be the class of all MV-algebras satisfying equations (E_{n0}), (E_{np}), for each integer p such that $2 \leq p < k$ and $p \notin \bigcup_{r \in I \cup J} \mathbf{Div}(r-1)$, and (F_{nq}), for each integer $q \geq 2$ such that $q \in \bigcup_{r \in I} (\mathbf{Div}(r-1) \setminus \bigcup_{s \in J} \mathbf{Div}(s-1))$. It follows from Lemmas 8.4.1 and 8.4.5, that $K_m \in C$ if and only $m-1$ divides $s-1$ for some $s \in J$. Let $r \in I$ and $y \in Ł_{r+1}$. Since $y^q \in Ł_{r+1}$, by Theorem 8.2.2 we have $(r-1).y^q = r.y^q$, whence a fortiori, $n.y^q = (n+1).y^q$. Since for each m, $Ł_m$ is isomorphic to a subalgebra of K_m, by Lemma 8.4.5 we can now conclude that $Ł_m$ is a member of C if and only if $k-1$ divides $r-1$ for some $r \in I \cup J$. Hence by Theorem 8.4.4, $C = C(\{Ł_i\}_{i \in I}, \{K_j\}_{j \in J})$. □

8.5 Varieties generated by a finite chain

Corollary 8.2.4 gives a simple equational characterization of the varieties of MV-algebras generated by $Ł_2$ and $Ł_3$. The following theorem, which is an immediate consequence of Theorems 1.4.6 and 8.2.2 together with Lemma 8.4.5, gives a set of defining equations for each variety generated by a finite simple MV-algebra, in all remaining cases. These equations are simpler than those obtainable by letting $I = \{n\}$ and $J = \emptyset$ in Theorem 8.4.6.

Theorem 8.5.1 *For every integer* $n \geq 4$ *and every MV-algebra* A, *the following conditions are equivalent:*

(*i*) *A satisfies the equations*

$$(E_{n0}) \qquad (n-1).x = n.x$$

and

$$(E_{np}) \qquad (p.x^{p-1})^n = n.x^p.$$

for every integer $p = 2, ..., n-2$ *that does not divide* $n-1$;

(ii) A *is a product of subalgebras of* $\mathbf{Ł}_n$;

(iii) $A \in \mathcal{C}(\mathbf{Ł}_n)$. □

Definition 8.5.2 Algebras in the variety $\mathcal{C}(\mathbf{Ł}_n)$ are called *n-valued* MV-algebras, or, for short, MV_n-algebras .

Corollary 8.5.3 *Let* A *be an* MV*-algebra.*

(i) If A *is a finite* MV_n*-algebra, then* A *is isomorphic to the direct product of a family of subalgebras of* $\mathbf{Ł}_n$*;*

(ii) If A *is finite then there is an integer* $n \geq 2$ *such that* $A \in \mathcal{C}(\mathbf{Ł}_n)$*;*

(iii) If A *is isomorphic to the direct product of a finite family of subalgebras of* $\mathbf{Ł}_n$ *then the members of this family coincide with the images of* A *under all possible homomorphisms of* A *into* $\mathbf{Ł}_n$.

Proof: (i) By Proposition 3.6.5, A is isomorphic to the direct product of a finite family of finite chains. Since any such chain L is a homomorphic image of A, $L \in \mathcal{C}(\mathbf{Ł}_n)$. By Theorem 8.5.1(ii), L must be a subalgebra of $\mathbf{Ł}_n$.

(ii) By Proposition 3.6.5, we can write

$$A \cong \prod \mathbf{Ł}_{n_1}, \ldots, \mathbf{Ł}_{n_r}.$$

Let s be the least common multiple of $n_1 - 1, \ldots, n_r - 1$, and $n = s+1$. By the remark following Proposition 3.5.3 each of $\mathbf{Ł}_{n_1}, \ldots, \mathbf{Ł}_{n_r}$ is a subalgebra of $\mathbf{Ł}_n$. Thus, $A \in \mathcal{C}(\mathbf{Ł}_n)$.

(iii) Let J be an ideal of A. Then a fortiori J is an ideal of the underlying lattice $\mathbf{L}(A)$. By our finiteness assumption, there is $a \in A$ such that $J = (a] = \{x \in A \mid x \leq a\}$. By Corollary 1.5.6, $a \in \mathbf{B}(A)$. It is easy to see that J is a maximal ideal of A iff there is an atom b of $\mathbf{B}(A)$ such that $J = (\neg b]$. Let $h : A \to S_1 \times \ldots \times S_r$ an arbitrary isomorphism, where $S_1, \ldots, S_r$ are all subalgebras of Ł_n. Let $p_1, \ldots, p_r$ be the compositions of the canonical projections of Ł_n^r onto Ł_n with h. Trivially, p_i is a homomorphism of A into Ł_n. Further, for each $i = 1, \ldots, r$, it is easy to see that $p_i(A) = S_i$. Thus, for each i there is an atom b_i in $\mathbf{B}(A)$ such that $\text{Ker}(p_i) = (\neg b_i]$. It follows that $(\neg b_1] \cap \ldots \cap (\neg b_r] = (0]$, i.e., $b_1 \vee \ldots \vee b_r = 1$. In conclusion, $b_1, \ldots, b_r$ exhaust all possible atoms of $\mathbf{B}(A)$, and $p_1, \ldots, p_r$ are all possible homomorphisms of A into Ł_n. □

An MV$_n$-algebra A, with a distinguished subset Y of elements, is said to be a *free MV$_n$-algebra over the (generating set) Y* , and is denoted by $Free_Y^n$, iff for every MV$_n$-algebra B and every function $f: Y \to B$, f can be uniquely extended to a homomorphism of A into B.

As in the case of free MV-algebras (cf. Section 3.1), there is no danger of ambiguity in calling $Free_\kappa^n$ *the* free MV$_n$-algebra over κ many generators.

Since for MV-terms p, q, $p^{\text{Ł}_n} = q^{\text{Ł}_n}$ iff $p^A = q^A$ for each MV$_n$-algebra A, arguing as in the proof of Proposition 3.1.4 we obtain:

Proposition 8.5.4 *For each natural number $n \geq 2$ and cardinal $\kappa \geq 1$, $Free_\kappa^n = Term(\text{Ł}_n, \kappa)$.* □

8.6 The cardinality of $Free_r^n$

When κ is a finite cardinal, say $\kappa = r \geq 1$, $\text{Ł}_n^{\text{Ł}_n^r}$ is finite, and has in fact n^{n^r} elements. Therefore by the above Proposition 8.5.4 together with Corollary 8.5.3, $Free_r^n$ is isomorphic to a direct product of a finite number of subalgebras of Ł_n. In this section, for each r and n we shall determine how many times each subalgebra of Ł_n appears in this direct product.

Let $Y = \{y_1, \ldots, y_r\}$ be a generating set of the free MV$_n$-algebra $Free_r^n$. There are n^r functions from Y into Ł_n, denoted by $e_1, \ldots, e_{n^r}$. Let S_i be the smallest subalgebra of Ł_n containing all $e_i(y_1), \ldots, e_i(y_r)$, $i = 1, 2, \ldots, n^r$. Letting $h_i = h_{e_i}$ be the unique homomorphism of

$Free_r^n$ into $Ł_n$ extending e_i, by Proposition 8.5.4, $h_1, \ldots, h_{n^r}$ is the list of all homomorphisms of $Free_r^n$ into $Ł_n$; further, $h_i(Free_r^n) = S_i$, for each $i = 1, \ldots, n^r$. Thus, by Corollary 8.5.3(iii) we have

$$Free_r^n = S_1 \times \ldots \times S_{n^r}.$$

For each $i = 1, \ldots, n^r$ and $j = 1, \ldots, r$, we define

$$e_i(y_j) = \frac{a_{ij}}{(n-1)}.$$

Recall that $\mathbf{Div}(n)$, $(n \geq 2)$ denotes the set of all divisors $d \geq 1$ of n. Given integers $n_1, \ldots, n_k \geq 1$, we shall denote by $\gcd(n_1, \ldots, n_k)$ their greatest common divisor. By Proposition 3.5.3 together with the remark following it, the set of subalgebras of $Ł_n$ coincides with the set of algebras $Ł_{d+1}$, where $d \in \mathbf{Div}(n-1)$. Obviously, a subalgebra S_i coincides with $Ł_2$ if and only if, for all $j = 1, \ldots, r$ either $a_{ij} = 0$ or $a_{ij} = n-1$. Further, for each $d < n-1$ we have $S_i = Ł_{d+1}$ iff $\gcd(a_{i1}, \ldots, a_{ir}, n-1) = (n-1)/d$. Since $e_1, \ldots, e_{n^r}$ lists all functions of $\{y_1, \ldots, y_r\}$ into $Ł_n$, for any given $d \in \mathbf{Div}(n-1)$ there is an i such that $a_{ij} = (n-1)/d$ for all $j = 1, \ldots, r$. Such i also has the property that $S_i = Ł_{d+1}$. Therefore, all subalgebras of $Ł_n$ are included among the S_i's, and we have

$$Free_r^n = \Pi_{d \in \mathbf{Div}(n-1)} Ł_{d+1}^{\alpha(n,r,d)},$$

where $\alpha(n,r,1) = 2^r$, and for each $d > 1, \alpha(n,r,d)$ is the number of r-tuples $(a_{i1}, \ldots, a_{ir}) \in \{0, 1, \ldots, n-1\}^r$ such that $\gcd(a_{i1}, \ldots, a_{ir}, n-1) = (n-1)/d$.

Let us fix $n \geq 3, r \geq 1$ and $1 \neq d \in \mathbf{Div}(n-1)$. In order to compute the number $\alpha(n,r,d)$, we shall first introduce some notations. For each a and b in $\mathbf{Div}(n-1)$, by writing

$$a \prec b$$

we understand that b *covers* a, in the divisibility order. Stated otherwise, a divides b and there is no c with $c \neq a$ and $c \neq b$, such that a divides c and c divides b. For each $c \in \mathbf{Div}(n-1)$ we define

$$\mathbf{MaxDiv}(c) = \{a \in \mathbf{Div}(c) \mid a \prec c\},$$

$$\mathbf{Min}(c, n-1) = \{a \in \mathbf{Div}(n-1) \mid c \prec a\},$$

and

$$\mathbf{H}(c) = \{i \in \{1, \ldots, n^r\} \mid c \in \mathbf{Div}(a_{i1}) \cap \ldots \cap \mathbf{Div}(a_{ir})\}.$$

Let us note that $\mathbf{MaxDiv}(c) = \emptyset$ iff $c = 1$; moreover, $\mathbf{Min}(c, n-1) = \emptyset$ iff $c = n-1$. Further, $i \in \mathbf{H}(c)$ iff there are integers $0 \leq b_1, \ldots, b_r$ such that $b_j c = a_{ij} \leq n-1$,for all $j = 1, \ldots, r$. It follows that $\sharp\mathbf{H}(c) = (\frac{n-1}{c} + 1)^r$. We then have

$$\{i \in \{1, \ldots, n^r\} \mid \gcd(a_{i1}, \ldots, a_{ir},\ n-1) = \frac{n-1}{d}\}$$

$$= \mathbf{H}(\frac{n-1}{d}) \setminus (\bigcup\{\mathbf{H}(c) \mid c \in \mathbf{Min}(\frac{n-1}{d},\ n-1)\}).$$

As a consequence, we can also write

$$\alpha(n, r, d) = \sharp\mathbf{H}(\frac{n-1}{d}) - \sharp\bigcup\{\mathbf{H}(c) \mid c \in \mathbf{Min}(\frac{n-1}{d},\ n-1)\}$$

$$= \sum(-1)^{\sharp X} \bigcap_{c \in X} \mathbf{H}(c),$$

where

$$\emptyset \neq X \subseteq \mathbf{Min}(\frac{n-1}{d}, n-1).$$

For each nonempty $X \subseteq \mathbf{Min}(\frac{n-1}{d}, n-1)$ let $\mathrm{lcm}(X)$ be the least common multiple of the elements of X. Then $\bigcap_{c \in X} \mathbf{H}(c) = \mathbf{H}(\mathrm{lcm}(X))$, whence

$$\sharp \bigcap_{c \in X} \mathbf{H}(c) = (\frac{n-1}{\mathrm{lcm}(X)} + 1)^r.$$

Note that $a \in \mathbf{Min}(\frac{n-1}{c}, n-1)$ iff $\frac{n-1}{a} \in \mathbf{MaxDiv}(c)$. Defining now

$$X^* = \{\frac{n-1}{c} \mid c \in X\},$$

we get $X \subseteq \mathbf{Min}(\frac{n-1}{d}, n-1)$ iff $X^* \subseteq \mathbf{MaxDiv}(d)$. We have just proved that $\frac{n-1}{\mathrm{lcm}(X)} = \gcd(X^*)$, for all $X \neq \emptyset$. Let us stipulate that $\gcd(\emptyset) = d$.

Theorem 8.6.1 *For any two integers $n \geq 2$ and $r \geq 1$ we have*

$$Free_r^n = \prod_{d \in \mathbf{Div}(n-1)} \mathbf{Ł}_{d+1}^{\alpha(n,r,d)},$$

where

$$\alpha(n,r,d) = \sum_{X \subseteq \mathbf{MaxDiv}(d)} (-1)^{\sharp X}(\gcd(X)+1)^r. \ \Box$$

From the above theorem, we get in particular $Free_r^n = \mathbf{Ł}_2^{2^r}$, which is a well known characterization of the free boolean algebra with r free generators. Further, whenever $n-1$ is prime we get

(8.11) $Free_r^n = \mathbf{Ł}_2^{2^r} \times \mathbf{Ł}_n^{n^r - 2^r}$.

A simple counting argument shows that $\sharp Free_1^2 = 4 =$ number of functions from $\mathbf{Ł}_2$ into $\mathbf{Ł}_2$.

As another example, while the number of functions from $\mathbf{Ł}_3$ into $\mathbf{Ł}_3$ is 27, a simple computation shows that $\sharp Free_1^3 = 12$.

In the rest of this section we shall prove some interesting and less immediate consequences of Theorem 8.6.1.

Let π be an MV-term in the variables $x_1, \ldots, x_r$, and suppose that S is a subalgebra of an MV-algebra A. From the definitions given in Section 1.4, it follows that for all $a_1, \ldots, a_r \in S$

$$\pi^A(a_1, \ldots, a_r) = \pi^S(a_1, \ldots, a_r) \in S.$$

Therefore, $\pi^A(S^r) \subseteq S$. For short, we say that *term functions preserve subalgebras.*

Since the subalgebras of $\mathbf{Ł}_n$ are precisely the algebras $\mathbf{Ł}_{d+1}$, for $d \in \mathbf{Div}(n-1)$, we can say that a function $f: \mathbf{Ł}_n^r \to \mathbf{Ł}_n$ preserves subalgebras if and only if f satisfies the following condition:

($\star$) *For each r-tuple $(\frac{a_1}{n-1}, \ldots, \frac{a_r}{n-1}) \in \mathbf{Ł}_n^r$, if $f(\frac{a_1}{n-1}, \ldots, \frac{a_r}{n-1}) = \frac{b}{n-1}$, then b is a multiple of* $\gcd(a_1, \ldots, a_r, n-1)$.

Conversely, the next result shows that if a function $f: \mathbf{Ł}_n^r \to \mathbf{Ł}_n$ preserves subalgebras, then f is a term function. More precisely we have

Corollary 8.6.2 *Given any arbitrary integers $n \geq 2$ and $r \geq 1$, let $f: \text{Ł}_n^r \to \text{Ł}_n$. Then $f \in Term(\text{Ł}_n, r)$ iff f preserves subalgebras.*

Proof: Let $2 \leq d_1 < d_2 < \ldots < d_s = n-1$ display the divisors of $n-1$ other than 1. Let $(b_{i1}, \ldots, b_{ir}), 1 \leq i \leq n^r$ display all r-tuples in Ł_n^r, where the indexes i are so chosen that

- precedence is given to the 2^r r-tuples that generate the subalgebra Ł_2 of Ł_n,
- followed by the $\alpha(n, r, d_2)$ many r-tuples that generate the subalgebra Ł_{d_2+1},
- and so on.

Let D be the subalgebra of Ł_n^r given by those functions satisfying condition $(\star)$. Let E be the product algebra

$$\Pi_{1 \leq i \leq s} \text{Ł}_{d_i}^{\alpha(n,r,d_i)} \cong Term(\text{Ł}_n, r).$$

It follows that the correspondence

$$f \mapsto (f(b_{i1}, \ldots, b_{1r}), \ldots, f(b_{n^r1}, \ldots, b_{n^rr}))$$

is an isomorphism of D onto E. □

8.7 Bibliographical remarks

The study of varieties of algebras is one of the main concerns of universal algebra. See, for instance, [104] or [150]. Proposition 8.1.1 originates from the early logico-algebraic Polish School (see [139]). Proposition 8.1.2 is due to Tarski [139, Theorem 20].

The notion of rank was introduced by Komori [130], who also proved the main results of Section 3. To prove Theorem 8.3.1 Komori introduced a special class of ordered abelian groups. The proof given here, based on the properties of the variety of ℓ-groups is borrowed from [50].

After Komori's classification of all varieties of MV-algebras, considerable work has been done, by Gispert, Torrens, Panti and others, on their axiomatization, and on the classification and axiomatization of classes generated by MV-chains (see, e.g., [94], [95] [67], [68], [199]). The present equational characterization of all proper varieties of MV-algebras is due to Di Nola and Lettieri. MV_n-algebras were introduced by Grigolia [105], [106], who also gave their equational characterization. See also [212]. The variety generated by K_2 was considered in [7] and [65].

The description of the free MV_n-algebra given in Theorem 8.6.1 was announced by António Monteiro in the *VI Congress of Mathematicians of Latin Expression*, held in Bucharest during September 1969.

The case $n = 3$ of equation (8.11) was established by A. Monteiro in his lectures delivered at the Universidad Nacional del Sur, Bahía Blanca, in 1963. The case $n = 4$ was first obtained in [44].

Corollary 8.6.2 was first proved by McNaughton in [152]. Our proof is obtained by means of direct techniques from universal algebra.

A brute force method to establish the nonisomorphism of finite algebras is to count their elements. For instance, as proved in [44], the 5-valued Moisil-Łukasiewicz algebra over one free generator has 192 elements, while, by (8.11), the free MV_5-algebra over one generator has 300 elements.

Chapter 9

Advanced topics

The first part of this chapter deals with disjunctive normal forms in the infinite-valued calculus of Łukasiewicz. We shall generalize the Farey-Schauder machinery of Chapter 3 to formulas in any number of variables. Disjunctive normal forms will be the key tool to prove McNaughton's theorem, generalizing the proof given in 3.2.8 for functions of one variable. We shall also discuss the relationships between normal form reductions and toric desingularizations, and the correspondence between MV-algebras and AF C^*-algebras. Strengthening Corollary 4.5.3, we shall show that the tautology problem in the infinite-valued calculus is in fact co-NP-complete, thus having the same complexity as it boolean counterpart. We shall give a proof of Di Nola's representation theorem for all MV-algebras.

With the possible exception of the first section, the pace in this chapter is usually faster than in the previous chapters. Prerequisites ranging from ℓ-group theory, polyhedral topology, NP-completeness theory, algebraic geometry, functional analysis, model theory, may be necessary for a complete understanding of the results in the following sections. In a final section appropriate references shall be given to the interested readers.

9.1 McNaughton's theorem

For each (closed) n-dimensional simplex $S \subseteq [0,1]^n$ with rational vertices we let

(I) $\mathbf{v}_0, \mathbf{v}_1, \ldots, \mathbf{v}_n$ be the list of vertices of S;

(II) $1 \leq d_i =$ least common denominator of the coordinates of $\mathbf{v}_i$;

(III) v_{ij} be the uniquely determined integers satisfying

$$(9.1)\quad \mathbf{v}_i = (v_{i0}/d_i, \ldots, v_{i(n-1)}/d_i),\ 0 \leq v_{ij} \leq d_i,\ (j = 0, \ldots, n-1)$$

with $\gcd(v_{i0}, \ldots, v_{i(n-1)}, d_i) = 1$, where, as in the previous chapters, *gcd* denotes greatest common divisor;

(IV) $\mathbf{v}_i^{hom} \in \mathbf{Z}^{n+1}$ be the *homogeneous* coordinates of the vertices of S,

$$(9.2)\quad \mathbf{v}_i^{hom} = (v_{i0}, \ldots, v_{i(n-1)}, d_i),\quad i = 0, 1, \ldots, n;$$

(V) $\mathcal{M}_S$ be the $(n+1) \times (n+1)$ matrix whose ith row coincides with $\mathbf{v}_i^{hom}$. We say that the list of vertices $\mathbf{v}_0, \mathbf{v}_1, \ldots, \mathbf{v}_n$ has the *positive orientation* iff $1 \leq \det(\mathcal{M}_S)$.

Following tradition, the set

$$(9.3)\quad \mathcal{P}_S = \{\mathbf{z} \in \mathbf{R}^{n+1} \mid \mathbf{z} = \lambda_0 \mathbf{v}_0^{hom} + \ldots + \lambda_n \mathbf{v}_n^{hom},\ 0 \leq \lambda_i < 1,\ \lambda_i \in \mathbf{R}\}$$

shall be called the *half-open parallelepiped* of S.

Definition 9.1.1 Adopt the above notation and terminology. Then S is said to be a *unimodular* simplex iff $\det(\mathcal{M}_S) = 1$. By a *unimodular triangulation* of $[0,1]^n$ we mean a set $\mathcal{W}$ of closed n-dimensional unimodular simplexes with rational vertices, such that the union of all simplexes in $\mathcal{W}$ coincides with $[0,1]^n$, and any two simplexes in $\mathcal{W}$ are either disjoint or intersect in a common face.

The following result strengthens Proposition 3.3.1:

Theorem 9.1.2 *Let $f : [0,1]^n \to [0,1]$ be a McNaughton function, with its linear constituents $p_1, \ldots, p_k$. Then there is a* unimodular *triangulation $\mathcal{U}$ of $[0,1]^n$ such that for each simplex $S \in \mathcal{U}$, f coincides with some p_j over S.*

Proof: Let $\mathcal{S}$ be as given by Proposition 3.3.1. Adopting the above notation, let T be an n-dimensional simplex in $\mathcal{S}$ such that $\det(\mathcal{M}_T)$ is maximum. Let $\mathbf{v}_0, \ldots, \mathbf{v}_n$ be the positively oriented list of vertices of T. If $\det(\mathcal{M}_T) = 1$ we are done. Otherwise, $\det(\mathcal{M}_T) > 1$ and $(\mathcal{M}_T)^{-1}$ is not an integral matrix.

Let $\mathbf{e}_0, \ldots, \mathbf{e}_n$ be the standard basis vectors of $\mathbf{R}^{n+1}$: thus, $\mathbf{e}_0 = (1,0,0,\ldots,0)$, $\mathbf{e}_1 = (0,1,0,\ldots,0)$, $\ldots$, $\mathbf{e}_n = (0,0,0,\ldots,1)$. We can safely assume that at least one of the coordinates of $\mathbf{e}_0$ in the basis given by the vectors $\mathbf{v}_0^{hom}, \ldots, \mathbf{v}_n^{hom}$, is not an integer. Among all vectors that are obtainable by adding to $\mathbf{e}_0$ linear combinations with integral coefficients of the vectors $\mathbf{v}_i^{hom}$, there is precisely one, denoted $\mathbf{w} \in \mathbf{R}^{n+1}$, (*read:* $\mathbf{w} = \mathbf{e}_0$ modulo $\mathcal{P}_T$), lying in the half-open parallelepiped $\mathcal{P}_T$. We have shown

$$0 \neq \mathbf{w} = (w_0, \ldots, w_{(n-1)}, w_n) \in \mathbf{Z}^{n+1} \cap \mathcal{P}_T.$$

Let $\mathbf{v}_T$ be the *inhomogeneous correspondent* of $\mathbf{w}$, in symbols, $\mathbf{v}_T = (w_0/d, \ldots, w_{(n-1)}/d)$, where $d = w_n =$ last coordinate of $\mathbf{w}$. Trivially, $\mathbf{v}_T$ is a point of the n-simplex T, and is not one of its vertices. For any $(n-1)$-dimensional face F of T not containing $\mathbf{v}_T$, the point $\mathbf{v}_T$ together with the vertices of F determines a new n-dimensional simplex $[\mathbf{v}_T, F] \subseteq T$. Let $\mathcal{M} = \mathcal{M}_{[\mathbf{v}_T, F]}$ be the associated matrix. Since the nonzero integral vector $\mathbf{w}$ belongs to the half-open parallelepiped $\mathcal{P}_T$, by (9.3), the determinant of $\mathcal{M}$ is an integer satisfying the inequalities

$$(9.4)\quad 1 \leq \det(\mathcal{M}) < \det(\mathcal{M}_T).$$

Let $\mathcal{E}$ be the set of all n-dimensional simplexes of the form $[\mathbf{v}_T, F]$, where F ranges over the $(n-1)$-dimensional faces of T not containing the point $\mathbf{v}_T$. Replacing in $\mathcal{S}$ the simplex T by the new simplexes of $\mathcal{E}$, we obtain a *rational subdivision* of $\mathcal{S}$, i.e., a set $\mathcal{S}_1$ of n-dimensional simplexes with rational vertices such that

(i) each simplex of $\mathcal{S}$ is a union of simplexes of $\mathcal{S}_1$, and

(ii) any two simplexes in $\mathcal{S}_1$ are either disjoint or they intersect in a common face.

The set of vertices of (simplexes in) $\mathcal{S}_1$ is the same as in $\mathcal{S}$, with the only addition of $\mathbf{v}_T$. By (9.4), the number of simplexes R in $\mathcal{S}_1$ such that $\det(\mathcal{M}_R) = \det(\mathcal{M}_T)$ is decreased. In this way, proceeding by induction, we obtain a sequence of sets of n-dimensional simplexes

$$\mathcal{S} = \mathcal{S}_0, \quad \mathcal{S}_1, \quad \mathcal{S}_2, \ldots.$$

This sequence must terminate after a finite number z of steps, yielding a set $\mathcal{S}_z = \mathcal{U}$ of unimodular n-dimensional simplexes having the required properties. □

Definition 9.1.3 Given a unimodular triangulation $\mathcal{W}$ of $[0,1]^n$, and a vertex $\mathbf{v}$ of some simplex in $\mathcal{W}$, let d be the least common denominator of the coordinates of $\mathbf{v}$. Then the *Schauder hat* $h_{\mathbf{v}}$ of $\mathcal{W}$ at $\mathbf{v}$ is the continuous function $h_{\mathbf{v}} : [0,1]^n \to [0,1]$ determined by the following conditions:
(i) $h_{\mathbf{v}}(\mathbf{v}) = 1/d$;
(ii) $h_{\mathbf{v}}(\mathbf{u}) = 0$ for each vertex $\mathbf{u} \in \mathcal{W}$ different from $\mathbf{v}$;
(iii) $h_{\mathbf{v}}$ is linear over each simplex $U \in \mathcal{W}$.

We denote by $\mathcal{H}_{\mathcal{W}}$ the set of Schauder hats of $\mathcal{W}$.
The *star* of $\mathbf{v}$, in the triangulation $\mathcal{W}$—in symbols, $\mathrm{star}(\mathbf{v})$—is the set of all n-dimensional simplexes of $\mathcal{W}$ having $\mathbf{v}$ among their vertices.

Proposition 9.1.4 *Let $\mathcal{W}$ be a unimodular triangulation of* $[0,1]^n$, $\mathbf{v}$ *a vertex in $\mathcal{W}$, and $h_{\mathbf{v}}$ be its corresponding Schauder hat. Then we have*
(i) For any simplex $S \in \mathrm{star}(\mathbf{v})$, let g be the linear polynomial coinciding with $h_{\mathbf{v}}$ over S; then there are integers $a_0, \ldots, a_{(n-1)}, b$ such that for every $\mathbf{x} = (x_0, \ldots, x_{(n-1)}) \in S$,

$$(9.5) \quad h_{\mathbf{v}}(x_0, \ldots, x_{(n-1)}) = a_0 x_0 + \ldots + a_{(n-1)} x_{(n-1)} + b = g(\mathbf{x});$$

(ii) $h_{\mathbf{v}} \in Free_n$.

Proof: (i) Let $\mathbf{w}_0, \ldots, \mathbf{w}_n$ be a positively oriented list of vertices of S. By hypothesis, $\mathbf{v} = \mathbf{w}_l$, for some $l = 0, \ldots, n$. Then the sequence of coefficients $a_0, \ldots, a_{(n-1)}, b$ coincides with the lth column of the inverse matrix $(\mathcal{M}_S)^{-1}$; the latter is an integral matrix, by the unimodularity of $\mathcal{W}$.

(ii) Let $g_1, \ldots, g_q$ be the distinct linear polynomials given by (i). Our aim is to show that $h_{\mathbf{v}}$ coincides, over $\bigcup \mathrm{star}(\mathbf{v})$, with a $(\bigvee \bigwedge)$-combination of the g_i's. To this purpose, for any permutation σ of the set $\{1,\ldots,q\}$, we define the closed convex polyhedron P_σ by

(9.6) $$P_\sigma = \{\mathbf{x} \in [0,1]^n \mid g_{\sigma(1)}(\mathbf{x}) \geq g_{\sigma(2)}(\mathbf{x}) \geq \ldots \geq g_{\sigma(q)}(\mathbf{x})\}.$$

Let Ω be the set of permutations π such that P_π is n-dimensional. Then

(9.7) $$\mathbf{v} \in \bigcap_{\pi\in\Omega} P_\pi, \quad [0,1]^n = \bigcup_{\pi\in\Omega} P_\pi.$$

For any $\mu, \nu \in \Omega$ the polyhedra P_μ and P_ν are either disjoint, or else they intersect in a common face.

Fix now an arbitrary permutation $\pi \in \Omega$. Then, however we choose a permutation $\rho \in \Omega$ and $i \neq j$, the difference $g_{\pi(i)} - g_{\pi(j)}$ will never vanish in the interior of P_ρ. Moreover, there is a unique index $i_\pi \in \{1,\ldots,q\}$ such that

(9.8) $$h_{\mathbf{v}} = g_{\pi(i_\pi)} \quad \text{over} \quad P_\pi \cap \bigcup \mathrm{star}(\mathbf{v}).$$

Again keeping π fixed, let us define the function $g^\pi : \mathbf{R}^n \to \mathbf{R}$ by

(9.9) $$g^\pi = g_{\pi(1)} \wedge \ldots \wedge g_{\pi(i_\pi)}.$$

Claim. $g^\pi \leq h_{\mathbf{v}}$ over each simplex $W \in \mathrm{star}(\mathbf{v})$.

It is sufficient to prove the claim for all $\mathbf{y}$ in a suitably small open n-dimensional ball B centered at $\mathbf{v}$. By continuity, we can safely assume $\mathbf{y}$ to lie in the interior of P_ρ for some permutation $\rho \in \Omega$. By (9.8) there is an index $i \in \{1,\ldots,q\}$ such that $h_{\mathbf{v}} = g_{\pi(i)}$ over $P_\rho \cap B$. If $i \leq i_\pi$, the claim trivially follows by definition of g^π. If $i > i_\pi$, there is a point $\mathbf{x} \neq \mathbf{y}$ in the interior of $P_\pi \cap B$ such that $g_{\pi(i)}(\mathbf{x}) < g_{\pi(i_\pi)}(\mathbf{x}) = h_{\mathbf{v}}(\mathbf{x})$. We may assume $\mathbf{x}$ to be so close to $\mathbf{v}$ that the line segment $[\mathbf{x},\mathbf{y}]$ joining $\mathbf{x}$ and $\mathbf{y}$ lies in the interior of $B \bigcap (\bigcup \mathrm{star}(\mathbf{v}))$. Let $X, Y \in \mathbf{R}^{n+1}$ be defined by $X = (\mathbf{x}, h_{\mathbf{v}}(\mathbf{x}))$ and $Y = (\mathbf{y}, h_{\mathbf{v}}(\mathbf{y}))$. Let $[X,Y]$ denote the line segment joining X and Y. Let η be the restriction of $h_{\mathbf{v}}$ to $[\mathbf{x},\mathbf{y}]$. Then η is a continuous function, consisting of finitely many linear pieces. By our assumptions about i and $\mathbf{y}$, there is $\mathbf{y}' \neq \mathbf{y}$ such that η coincides with $g_{\pi(i)}$ over the interval $N = [\mathbf{y}', \mathbf{y}] \subseteq [\mathbf{x}, \mathbf{y}]$. Thus by our assumption about $\mathbf{x}$, over the half-open interval $N \setminus \{\mathbf{y}\}$ the graph of η lies strictly below the segment $[X, Y]$. Among all $\mathbf{w} \in [\mathbf{x}, \mathbf{y}]$ different from $\mathbf{y}$ and such that $(\mathbf{w}, \eta(\mathbf{w})) \in [X, Y]$, there is a point $\mathbf{z}$ nearest to $\mathbf{y}$ (It is quite possible that $\mathbf{z} = \mathbf{x}$). Let $j \in \{1, \ldots, q\}$ be such that $\eta(\mathbf{z}) = g_{\pi(j)}(\mathbf{z})$, and η coincides with $g_{\pi(j)}$ on a small open interval $N' = [\mathbf{z}, \mathbf{z}'] \subseteq [\mathbf{z}, \mathbf{y}]$, for some $\mathbf{z}' \neq \mathbf{z}$. Then the restriction of the graph of $g_{\pi(j)}$ to the half-open interval $N' \setminus \{\mathbf{z}\}$ lies strictly below the segment $[X, Y]$. Moving from $\mathbf{z}$ to $\mathbf{y}$, we get $g_{\pi(j)}(\mathbf{y}) < h_{\mathbf{v}}(\mathbf{y})$. Moving from $\mathbf{z}$ to $\mathbf{x}$, we get $g_{\pi(j)}(\mathbf{x}) > h_{\mathbf{v}}(\mathbf{x}) = g_{\pi(i_\pi)}(\mathbf{x})$, whence $j < i_\pi$. Thus, $g^\pi \leq g_{\pi(j)}$ over all of $\mathbf{R}^n$, which settles our claim.

To conclude the proof, let $g^\Omega = \bigvee_{\pi \in \Omega} g^\pi$. By our claim, together with (9.8), since for each $\sigma \in \Omega$, $g^\Omega = g^\sigma = h_{\mathbf{v}}$ over $P_\sigma \cap \bigcup \mathrm{star}(\mathbf{v})$, then $g^\Omega = h_{\mathbf{v}}$ over star(v). Trivially, $g^\Omega \vee 0 = h_{\mathbf{v}} = 0$ over $[0,1]^n \setminus \bigcup \mathrm{star}(\mathbf{v})$. Since $g^\Omega \vee 0$ is a $(\bigvee \bigwedge)$-combination of the $g_i^\sharp$ and the latter, by (i) and Lemma 3.1.9, are elements of $Free_n$, the desired conclusion follows from Proposition 1.1.5. □

Theorem 9.1.5 *For each cardinal κ, the free MV-algebra $Free_\kappa$ is given by the McNaughton functions over $[0,1]^\kappa$, with pointwise operations.*

Proof: In the light of Propositions 3.1.4 and 3.1.8 it suffices to show that every McNaughton function $f : [0,1]^n \to [0,1]$ belongs to $Free_n$.

To this purpose, let $\mathcal{U}$ be as in Theorem 9.1.2. Let $\mathbf{u} \in \mathbf{Q}^n$ be an arbitrary vertex of $\mathcal{U}$, and let d be the least common denominator of the coordinates of $\mathbf{u}$. Then $f(\mathbf{u}) = m_{\mathbf{u}}/d$ for some integer $0 \leq m_{\mathbf{u}} \leq d$. Moreover, f is linear over each n-dimensional simplex of $\mathcal{U}$. Let $h_{\mathbf{u}} \in \mathcal{H}_{\mathcal{U}}$ be as in Definition 9.1.3. Then the two continuous functions f and $\Sigma_{\mathbf{u}} m_{\mathbf{u}} h_{\mathbf{u}}$ coincide over all of $[0,1]^n$. Replacing sum by truncated sum $\oplus$, we have

$$(9.10) \quad f = \oplus_{\mathbf{u}}\, m_{\mathbf{u}}.h_{\mathbf{u}} = h_{\mathbf{u}} \oplus h_{\mathbf{u}} \oplus \ldots \oplus h_{\mathbf{u}} \quad (m_{\mathbf{u}} \text{ times}).$$

Since, by Proposition 9.1.4, $h_{\mathbf{u}} \in Free_n$, then $f \in Free_n$, a required. □

Remark: The set $\mathcal{H}_{\mathcal{U}}$ in the above proof is a DNF (*Disjunctive Normal Form*) reduction of f.

9.2 Nonsingular fans and normal forms

This section requires familiarity with the so called *vocabulary* from toric varieties to fans. Writing in homogeneous coordinates the vertices of each simplicial complex arising in the disjunctive normal form of a McNaughton function, one obtains a sequence leading to a nonsingular refinement of the *fan* (a fan being a complex of simplicial cones, as defined below) corresponding to the linear domains of the proposition. This refinement process in turn amounts to a resolution of singularities for toric variety corresponding to the fan.

In more detail, let $\mathcal{S}_0, \mathcal{S}_1, \ldots, \mathcal{S}_z = \mathcal{U}$ be a sequence of simplicial complexes over $[0,1]^n$, as in the proof of Theorem 9.1.2. Each simplex $T \in \mathcal{S}_i$ has vertices $\mathbf{v}_0, \ldots, \mathbf{v}_n$. The homogeneous counterpart of T is a *simplicial cone*, $\sigma^T = \langle \mathbf{v}_0{}^{hom}, \ldots, \mathbf{v}_n{}^{hom} \rangle$, i.e., the positive span in $\mathbf{R}^{n+1}$ of vectors $\mathbf{v}_0{}^{hom}, \ldots, \mathbf{v}_n{}^{hom}$. As a vector in $\mathbf{Z}^{n+1}$, each $\mathbf{v}_j{}^{hom}$ is *primitive*, i.e., minimal along its ray. The $\mathbf{v}_j{}^{hom}$ are called the primitive generating vectors of σ^T. From each $\mathcal{S}_i$ we obtain a *simplicial fan* Δ_i, i.e., a complex of simplicial cones. The fact that $\mathcal{S}_z$ is a unimodular triangulation is equivalent to saying that the primitive generating vectors of each $(n+1)$-dimensional cone in Δ_z form a unimodular matrix—for short, Δ_z is a *nonsingular* fan.

As is well known, every (nonsingular) fan Δ is canonically associated with a (nonsingular) toric variety X_Δ, in such a way that the sequence $\Delta_0, \ldots, \Delta_z$ corresponds to a desingularization X_z of the toric variety X_0. Thus, desingularizing a toric variety amounts to subdividing a simplicial complex into a unimodular triangulation, precisely as is done to compute DNF reductions of McNaughton functions. For the sake of definiteness, recalling Definition 9.1.3, let us give the following

Definition 9.2.1 A *Schauder set* in $[0,1]^n$ is a set of the form $\mathcal{H} = \mathcal{H}_\mathcal{U}$ for some (necessarily unique) unimodular triangulation of $[0,1]^n$. We say that $\mathcal{H}'$ is a *one-step star refinement* of $\mathcal{H}$ iff it is obtained from $\mathcal{H}$ as follows:

(a) Pick a set $\mathcal{S} = \{h_1, \ldots, h_q\} \subseteq \mathcal{H}$ and let $h_\mathcal{S} = h_1 \wedge \ldots \wedge h_q$;

(b) For each $j = 1, \ldots, q$ replace h_j by $h_j \ominus h_\mathcal{S}$;

(c) If $0 \neq h_\mathcal{S}$ put $h_\mathcal{S}$ in $\mathcal{H}'$.

Trivially, $\mathcal{H} = \mathcal{H}'$ if $h_S = 0$ or if S is a singleton. When S has two elements and $h_S \neq 0$ we say that the refinement is *binary*. Writing $\mathcal{H} = \mathcal{H}_\mathcal{U}$, and assuming $0 \neq h_S$, it follows that $\mathcal{H}' = \mathcal{H}_{\mathcal{U}'}$, where $\mathcal{U}'$ is the unimodular triangulation obtained by starring $\mathcal{U}$ at the mediant point determined by the face S of $\mathcal{U}$. We say that $\mathcal{H}^*$ is a *star refinement* of $\mathcal{H}$ iff it is obtained from $\mathcal{H}$ via a path $\mathcal{H} = \mathcal{H}_0, \mathcal{H}_1, \ldots, \mathcal{H}_t = \mathcal{H}^*$, where each $\mathcal{H}_i$ is a one-step star refinement of $\mathcal{H}_{i-1}$.

Theorem 9.2.2 *Any two Schauder sets $\mathcal{I}$ and $\mathcal{L}$ in $[0,1]^n$ have a common star refinement.*

Proof: Readers familiar with toric varieties will recognize this statement as a reformulation of the strong form of Oda's conjecture. While for several years only the one-dimensional case of the conjecture was known to be true, (by Danilov's decomposition theorem) it appears that Morelli has finally settled the conjecture in the affirmative for the general case. □

Thus, for any unimodular triangulation $\mathcal{U}$ we can explicitly construct $\mathcal{H}_\mathcal{U}$ starting from any Schauder set $\mathcal{H}_\mathcal{W}$, and then applying only one (deduction) rule, namely the one-step star refinement. This method is more efficient than the inductive procedure given by Lemma 3.1.9.

Theorem 9.2.3 *For any two Schauder sets $\mathcal{H}$ and $\mathcal{I}$ there is a star refinement $\mathcal{H}^*$ of $\mathcal{H}$ such that every element of $\mathcal{I}$ is a (truncated) sum of elements of $\mathcal{H}^*$, as in the above formula (9.10). Moreover, all one-step star refinements leading from $\mathcal{H}$ to $\mathcal{H}^*$ may be assumed to be binary.*

Proof: This is a consequence of the De Concini-Procesi theorem on elimination of points of indeterminacy in toric varieties. □

In the particular case when $n = 1$, using the Hirzebruch-Jung continued fraction algorithm, one can compute, for every fan Δ in the cartesian plane, the *coarsest* nonsingular subdivision of Δ. In MV-algebraic terms we have:

Corollary 9.2.4 *Every set $Q = \{f_1, \ldots, f_k\} \subseteq Free_1$ has a* least *DNF reduction $\mathcal{H}$, i.e., a Schauder set $\mathcal{H}$ satisfying the following conditions:*

- *(i) Each f_i is a (truncated) sum of the hats in $\mathcal{H}$ as in (9.10);*

- *(ii) Whenever a Schauder set $\mathcal{L}$ is a DNF reduction of Q, then $\mathcal{L}$ is also a DNF reduction of $\mathcal{H}$.*

□

9.3 Complexity of the tautology problem

In this section we require some familiarity with the theory of NP-completeness. We shall consider the following problem:

> INSTANCE: A formula ϕ.
>
> QUESTION: Is ϕ a tautology in the infinite-valued calculus of Łukasiewicz ?

We shall prove that the problem is co-NP-complete, i.e., the complementary problem of deciding whether a formula is not a tautology, is NP-complete. Cook's theorem states the co-NP-completeness of the tautology problem for the classical propositional calculus.

We shall denote by $|\phi|$ the number of occurrences of symbols in ϕ. We shall also use the notation $|x|$ for the absolute value of a real number x. This will never cause any confusion. Recalling Propositions 3.1.8 and 4.5.5, for each formula $\phi = \phi(X_1, \ldots, X_n)$ its associated McNaughton function shall be denoted by $f_\phi = f_\phi(x_1, \ldots, x_n)$. For all $\mathbf{x}, \mathbf{y} \in [0,1]^n$, the *one-sided derivative* of f_ϕ at $\mathbf{x}$ along direction $\mathbf{d} = \mathbf{y} - \mathbf{x}$ is defined by

$$f'_\phi(\mathbf{x}; \mathbf{d}) = \lim_{\epsilon \downarrow 0} \frac{f_\phi(\mathbf{x} + \epsilon\mathbf{d}) - f_\phi(\mathbf{x})}{\epsilon}.$$

Proposition 9.3.1 *With the above notation, the directional derivative $f'_\phi(\mathbf{x}; \mathbf{d})$ is well defined, and we have the inequality*

$$(9.11) \quad |f'_\phi(\mathbf{x}; \mathbf{d})| \leq ||\mathbf{d}|| \cdot |\phi|,$$

where $||\mathbf{d}||$ denotes euclidean norm in $\mathbf{R}^n$.

Proof: The existence of $f'_\phi(\mathbf{x}; \mathbf{d})$ follows by definition of McNaughton function. Inequality (9.11) is proved by induction on the number m of connectives occurring in ϕ. The basis $m = 0$ is trivial. For the induction step, if $\phi = \neg\psi$ for some formula ψ, then the desired conclusion immediately follows from the identity $f_\phi = 1 - f_\psi$. Finally, if $\phi = \psi \oplus \chi$ then, assuming without loss of generality that both ψ and χ have the same variables, the desired conclusion immediately follows by definition of truncated addition, upon noting that $f_\phi = f_\psi \oplus f_\chi$. □

Corollary 9.3.2 *Let $p(x_1, \ldots, x_n) = c + m_1 x + \cdots + m_n x_n$ be a linear polynomial with integer coefficients $c, m_1, \ldots, m_n$. Let $f_\phi(x_1, \ldots, x_n)$ be the McNaughton function associated to a formula ϕ. Suppose f_ϕ coincides with p over an n-simplex $\mathcal{T} \subseteq [0,1]^n$. Then we have*

(9.12) $\max(|m_1|, \ldots, |m_n|) \leq |\phi|$.

□

Proposition 9.3.3 *Let $f_\phi(x_1, \ldots, x_n)$ be the McNaughton function associated to a formula ϕ. Assume f_ϕ does not coincide with the zero function over $[0,1]^n$. Then there exists a point*

$$\mathbf{x} = (a_1/b, \ldots, a_n/b) \in [0,1]^n$$

with $a_i, b \in \mathbf{Z}$ and $0 \leq a_i \leq b$ $(i = 1, \ldots, n)$ such that $f_\phi(\mathbf{x}) > 0$ and $0 < b < 2^{(4|\phi|^2)}$.

Proof: By Proposition 3.3.1 there is a finite number of distinct polynomials $p_1, \ldots, p_m$ with integer coefficients, and a finite number of n-dimensional simplexes $\mathcal{D}_1, \ldots, \mathcal{D}_s$, whose union is $[0,1]^n$, and such that over each $\mathcal{D}_j$ the function f_ϕ coincides with some $p_{i(j)}$ ($i(j) = 1, \ldots, m$). Suppose without loss of generality, $\mathbf{x} \in \mathcal{D}_1$ and $f_\phi(\mathbf{x}) > 0$. Then we can safely assume $\mathbf{x}$ to be a vertex of $\mathcal{D}_1$. Therefore, the coordinates of $\mathbf{x}$ are all rational, say $\mathbf{x} = (a_1/b, \ldots, a_n/b)$, for suitable integers a_i and b with $0 \leq a_i \leq b$. Moreover, by the above corollary, $\mathbf{x}$ is the solution of a system of n linear equations in n unknowns, and each row has its coefficients $\leq 2|\phi|$. Since, trivially, $n \leq |\phi|$ then by Hadamard's inequality we conclude that the determinant Δ of this system satisfies the inequality

$$|\Delta| \leq (4n|\phi|^2)^{n/2} < 2^{(4|\phi|^2)}.$$

Since $b \leq |\Delta|$, the desired conclusion immediately follows. □

Theorem 9.3.4 *The tautology problem for the infinite-valued calculus is in the class co-NP.*

Proof: A nondeterministic procedure quickly deciding if a formula

$$\phi(X_1, \ldots, X_n)$$

is *not* a tautology is as follows: Firstly, applying Proposition 9.3.3 to the function $1 - f_\phi$, guess a rational point

$$\mathbf{x} = (a_1/b, \ldots, a_n/b) \in [0,1]^n$$

such that $f_\phi(\mathbf{x}) < 1$ and $0 < b < 2^{(4|\phi|^2)}$. Secondly, for the purpose of checking that $f_\phi(\mathbf{x}) < 1$, write each coordinate a_i/b as a pair of binary integers; let $[a_i]$ and $[b]$ denote the number of bits of a_i and b. Note that $[a_i] \leq [b] \leq 4|\phi|^2$ for all $i = 1, \ldots, n$; also note that, once $\mathbf{x}$ is written down as a sequence of pairs of binary numbers, its length $[\mathbf{x}]$ will satisfy the inequalities

$$[\mathbf{x}] \leq 1 + n(2 + [b] + \max_i [a_i]) \leq 1 + n(2 + 8|\phi|^2) \leq 11|\phi|^3.$$

Since the operations of negation and truncated addition do not increase denominators, for some polynomial $q : \mathbf{N} \to \mathbf{N}$ the value $f_\phi(\mathbf{x})$ can be computed by a deterministic Turing machine within a number of steps $\leq q(|\phi|)$. □

In order to prove that the tautology problem in the infinite-valued calculus is co-NP-hard, we prepare:

Definition 9.3.5 For each integer $n \geq 1$ and $t \geq 2$ we define the $[0,1]$-valued function $f_{n,t}$ by stipulating that for all $\mathbf{x} = (x_1, \ldots, x_n) \in [0,1]^n$

$$f_{n,t}(\mathbf{x}) = \underbrace{\{(x_1 \vee \neg x_1) \odot \ldots \odot (x_1 \vee \neg x_1)\}}_{t \text{ times}} \odot \ldots$$

$$\ldots \odot \underbrace{\{(x_n \vee \neg x_n) \odot \ldots \odot (x_n \vee \neg x_n)\}}_{t \text{ times}}.$$

Further, for each integer $i \geq 1$ the formulas ϕ_i, $\psi_{i,t}$, and $\rho_{n,t}$ are defined by

(i) $\phi_i = X_i \vee \neg X_i$;

(ii) $\psi_{i,t} = \phi_i \odot \ldots \odot \phi_i$ (t times);

(iii) $\rho_{n,t} = \psi_{1,t} \odot \ldots \odot \psi_{n,t}$.

As an immediate consequence of the definition we have $f_{n,t} = f_{\rho_{n,t}}$. A tedious but straightforward inspection yields the following

Lemma 9.3.6 *Fix an enumeration of the vertices of the cube* $[0,1]^n$. *Let* $\mathbf{v}_j$ *be the* j*th vertex* ($j = 1, \ldots, 2^n$). *Let*

$$\mathcal{E}_{j1}, \ldots, \mathcal{E}_{jn}$$

enumerate the edges of $[0,1]^n$ *adjacent to* $\mathbf{v}_j$. *For each* $i = 1, \ldots, n$ *and* $t \geq 2$ *let* $\mathbf{y}_{ji}$ *be the point lying on edge* $\mathcal{E}_{ji}$ *at a distance* $1/t$ *from* $\mathbf{v}_j$. *Let* $\mathcal{T}_j$ *be the* n*-simplex with vertices* $\mathbf{v}_j, \mathbf{y}_{j1}, \ldots, \mathbf{y}_{jn}$. *Then we have*

(i) $f_{n,t}(\mathbf{v}_j) = 1$;

(ii) $f_{n,t}(\mathbf{y}_{ji}) = 0$;

(iii) $f_{n,t}$ *is linear over each simplex* $\mathcal{T}_j$;

(iv) $f_{n,t}$ *vanishes in* $[0,1]^n$ *outside* $\bigcup_{j=1}^{2^n} \mathcal{T}_j$. □

Lemma 9.3.7 *Adopt the above notation. Let* $\phi = \phi(X_1, \ldots, X_n)$ *be a formula. Let* $t = |\phi|$, *and say without loss of generality,* $t \geq 2$. *Then* ϕ *is a tautology in the boolean calculus iff* $\rho_{n,t} \rightarrow \phi$ *(i.e.,* $\neg\rho_{n,t} \oplus \phi$*) is a tautology in the infinite-valued calculus.*

Proof: For the nontrivial direction, by definition of implication, we must prove

$$f_{n,t} \leq f_\phi.$$

With reference to Lemma 9.3.6(iv), the inequality holds over the set $[0,1]^n \setminus \bigcup_j \mathcal{T}_j$. So let us assume that for some $j = 1, \ldots, 2^n$, $f_{n,t}(\mathbf{x}) > f_\phi(\mathbf{x})$ with $\mathbf{x} \in \mathcal{T}_j$ (absurdum hypothesis). By continuity we can safely assume $\mathbf{x}$ to be in the interior of $\mathcal{T}_j$, whence in particular, $\mathbf{x} \neq \mathbf{v}_j$. Let

$$\mathbf{u} = \frac{\mathbf{x} - \mathbf{v}_j}{||\mathbf{x} - \mathbf{v}_j||}.$$

By our analysis, together with Proposition 9.3.1, for each point $\mathbf{y}$ lying in the interval $[\mathbf{v}_j, \mathbf{x}]$ we have

$$f'_{n,t}(\mathbf{y}; \mathbf{u}) \leq -t = -|\phi|.$$

Further,

$$f'_\phi(\mathbf{y}; \mathbf{u}) \geq -|\phi|.$$

Thus, for all $\mathbf{y} \in [\mathbf{v}_j, \mathbf{x}]$,

$$f'_{n,t}(\mathbf{y}; \mathbf{u}) \leq -|\phi| \leq f'_\phi(\mathbf{y}; \mathbf{u}).$$

Since by assumption, ϕ is a boolean tautology, $f_\phi = 1$ on $\{0,1\}^n$. By Lemma 9.3.6,

$$f_{n,t}(\mathbf{v}_j) = f_\phi(\mathbf{v}_j) = 1.$$

Now, $f_{n,t}$ is linear on the interval $[\mathbf{v}_j, \mathbf{x}]$; on the other hand, f_ϕ is (continuous and) piecewise-linear on $[\mathbf{v}_j, \mathbf{x}]$. Thus $f_\phi \geq f_{n,t}$ on $[\mathbf{v}_j, \mathbf{x}]$, a contradiction. □

Theorem 9.3.8 *The tautology problem in Łukasiewicz infinite-valued calculus is co-NP-complete.*

Proof: We have just given a polynomial-time reduction of the boolean tautology problem into the tautology problem for the infinite-valued calculus. The desired conclusion is now an immediate consequence of Theorem 9.3.4 and the above Lemma 9.3.7, in the light of Cook's NP-completeness theorem for the boolean satisfiability problem. □

9.4 MV-algebras and AF C*-algebras

This section requires some familiarity with AF C^*-algebras. Every C^*-algebra $\mathcal{A}$ considered in this paper shall have a unit element $1_\mathcal{A}$. By a *projection* p in $\mathcal{A}$ we mean a self-adjoint idempotent $p = p^* = p^2$. Up to isomorphism, the most general possible finite-dimensional C^*-algebra $\mathcal{F}$ is a finite direct sum $\mathcal{M}_{d(1)} + \mathcal{M}_{d(2)} + \cdots + \mathcal{M}_{d(t)}$ where $\mathcal{M}_{d(i)}$ denotes the C^*-algebra of all $d(i) \times d(i)$ complex matrices, for suitable $1 \leq d(i)$.

An *approximately finite-dimensional* (for short, *AF C^*-algebra*) is the norm closure of the union of a sequence $\mathcal{F}_1 \subseteq \mathcal{F}_2 \subseteq \ldots$ of finite-dimensional C^*-algebras, all with the same unit, where each $\mathcal{F}_i$ is a $*$-subalgebra of $\mathcal{F}_{i+1}$.

For every AF C^*-algebra $\mathcal{A}$, two projections $p, q \in \mathcal{A}$ are said to be *equivalent* iff there exists an element $v \in \mathcal{A}$ such that $vv^* = p$ and $v^*v = q$. We denote by $[p]$ the equivalence class of p, and by $L(\mathcal{A})$ the set of equivalence classes of projections of $\mathcal{A}$. The *Murray-von Neumann order* over $L(\mathcal{A})$ is defined by

$$[p] \leq [q] \text{ iff } p \text{ is equivalent to a projection } r \text{ such that } rq = r.$$

Elliott's partial addition, denoted $+$, is the partial operation on $L(\mathcal{A})$ given by adding two projections whenever they are orthogonal. Then $+$ is associative, commutative, monotone, and satisfies the following *residuation* property:

(*) For every projection $p \in \mathcal{A}$, among all classes $[q]$ such that $[p] + [q] = [1_{\mathcal{A}}]$ there is a smallest one, denoted $\neg[p]$, namely the class $[1_{\mathcal{A}} - p]$.

Theorem 9.4.1 *For every AF C^*-algebra $\mathcal{A}$ we have:*
(i) There is at most one *extension of Elliott's partial addition to an associative, commutative, monotone operation $\oplus$ over the whole $L(\mathcal{A})$ having the above residuation property $(*)$. Such extension $\oplus$ exists iff $L(\mathcal{A})$ is a lattice;*

(ii) Let $\mathcal{K}(\mathcal{A}) = (L(\mathcal{A}), [0], \neg, \oplus)$. Then the map $\mathcal{A} \mapsto \mathcal{K}(\mathcal{A})$ is a one-one correspondence between isomorphism classes of AF C^*-algebras whose $L(\mathcal{A})$ is lattice-ordered, and isomorphism classes of countable MV-algebras;

(iii) In particular, up to isomorphism, the map $\mathcal{A} \mapsto \mathcal{K}(\mathcal{A})$ is a one-one correspondence between commutative AF C^*-algebras and countable Boolean algebras. The inverse correspondence is given by the map $X \mapsto C(X)$, the latter denoting the C^*-algebra of complex-valued continuous functions over an arbitrary separable totally disconnected compact Hausdorff space X. □

Remarks.

1. Classes of AF C^*-algebras $\mathcal{A}$ whose $L(\mathcal{A})$ is a lattice include commutative, finite-dimensional, continuous trace, liminary with Hausdorff spectrum, as well as all AF C^*-algebras with comparability of projections in the sense of Murray-von Neumann.

2. Intuitively, part (iii) in the above theorem suggests that MV-algebras are a noncommutative generalization of boolean algebras. Since, by Theorem 4.6.9, every countable MV-algebra is the Lindenbaum algebra of a theory Θ in the infinite-valued calculus of Łukasiewicz with denumerably many variables, any set of axioms for Θ is a presentation of *a unique* AF C^*-algebra $\mathcal{A}_\Theta$. The complexity of the word problem (in the sense of Theorem 4.6.10) of Θ is a faithful measure of the combinatorial complexity of $\mathcal{A}_\Theta$. While most AF C^*-algebras existing in the literature have polynomial time complexity, if the word problem Θ happens to be Gödel incomplete, then necessarily $\mathcal{A}_\Theta$ has a nontrivial ideal. This shows the incompatibility of two equally imprecise and interesting conjectures: (a) that the C^*-algebraic mathematizations of physical systems existing in nature should have no quotient structures, and (b) that Gödel incomplete AF C^*-algebras might exist in nature.

3. Readers familiar with Grothendieck's group will recognize in the ℓ-group $\mathcal{G}(\mathcal{K}(\mathcal{A}))$ the group $K_0(\mathcal{A})$ equipped with the order induced by the image $K_0(\mathcal{A})^+$ of the generating monoid of $K_0(\mathcal{A})$.

9.5 Di Nola's representation theorem

This section requires familiarity with model theory.

Theorem 9.5.1 *Up to isomorphism, every MV-algebra B is an algebra of $[0,1]^*$-valued functions over some set, where $[0,1]^*$ is an ultrapower of $[0,1]$, only depending on the cardinality of B.*

Proof: Let $\mathcal{P}(B)$ be the set of prime ideals of B. In the light of the Subdirect Representation Theorem 1.3.3, let us embed B into the MV-algebra $\prod\{B/I \mid I \in \mathcal{P}(B)\}$. For each prime ideal I of B, in the light of Theorem 7.1.7, let $G(I)$ be the totally ordered abelian group with strong unit $u(I)$ uniquely determined by the stipulation

$\Gamma(G(I), u(I)) \cong B/I$. Let us embed $G(I)$ into a totally ordered *divisible* abelian group $K(I)$ with the same strong unit $u(I)$. Let $D(I) = \Gamma(K(I), u(I))$. Then from Lemma 7.2.1 it follows that B/I is embeddable into $D(I)$. Since any totally ordered divisible abelian group is elementarily equivalent to the additive group $\mathbf{R}$ of real numbers with natural order, it follows that $D(I)$ is elementarily equivalent to the MV-algebra $[0,1]$. By *Frayne's theorem* in model theory, each $D(I)$ is elementarily embeddable in a suitable ultrapower $[0,1]^{*I}$ of $[0,1]$. The *joint embedding property* of first-order logic now yields an ultrapower $[0,1]^*$ (only depending on the cardinality of B), such that each MV-algebra $[0,1]^{*I}$ is elementarily embeddable into $[0,1]^*$. Thus every B/I is embeddable into $[0,1]^*$, whence the desired conclusion immediately follows. □

9.6 Bibliographical remarks

McNaughton's representation theorem 9.1.5 was first proved in [152], a few years before the advent of MV-algebras. The present proof, using unimodular triangulations and Schauder hats, was first given in [181]. Schauder hats over n-dimensional unimodular triangulations were introduced in [170]. Generalized Schauder hats are a standard tool in the theory of Banach spaces.

Ewald's book [82] gives a full account of the vocabulary between fans and toric varieties. Also see [195], where toric desingularizations are discussed, along with the various results of Danilov, De Concini-Procesi and Hirzebruch-Jung mentioned in this chapter. See Morelli [162] for a solution of the strong form of Oda's conjecture.

The relationship between toric desingularizations and the infinite-valued calculus of Łukasiewicz was first investigated in [183]. Direct applications to three-dimensional toric varieties are given in [5].

Theorem 9.3.8 was first proved in [167]. Any textbook on computational complexity theory contains a proof of Cook's NP-completeness theorem. A standard reference is [88].

See [77] and [101] for Elliott's theory, and for the role of Grothendieck's functor K_0 in connecting AF C^*-algebras and partially ordered abelian groups. AF C^*-algebras are a standard tool to give a mathematical description of spin systems in quantum statistical mechanics (see the references in [178]). The mutual relations between MV-algebras and AF C^*-algebras were first investigated in [163]. Theorem 9.4.1 was proved in [190] building on [163]. Gödel incompleteness phenomena for AF C^*-algebras are discussed in [163]. The fact that many well known examples of AF C^*-algebras are coded by polynomial-time theories in the infinite-valued calculus is noted in [166] and [177].

Theorem 9.5.1 is due to Di Nola, [63], [64]. The model theoretic machinery needed for its proof can be found in [42] and [158].

Chapter 10

Further Readings

10.1 More than two truth values

Łukasiewicz introduced many-valued logics in 1920. The history of studies of Łukasiewicz's original philosophical ideas and motivations is fairly long, and is definitely outside the scope of this book. Interested readers are referred to Wolenski's monograph [246], where Łukasiewicz's motivations are analyzed and his work on many-valuedness is presented in a wide perspective. In her essay [73] the author discusses many-valuedness in the framework on nonclassical logics. In their essay [204], Priest and Routley study Łukasiewicz logic from the viewpoint of paraconsistency. Paztig [200] discusses the relations between Łukasiewicz's ideas on many-valuedness and ideas in Chapter 9 of Aristotle's *De Interpretatione.* The short paper by Rosser [218] surveys the early stages of many-valued logic, and offers succinct historical and bibliographical remarks to an intended audience of physicists. The books [149], [30] and [227] contain English translations of papers by Łukasiewicz and Wajsberg.

The problem of interpreting nonboolean truth values was considered by many people, including Łukasiewicz. Prior to the interpretation via Ulam game, a game theoretic interpretation of infinite-valued logic was given by Giles in [93], in the context of subjective probability. While no probability is involved in Ulam game, it is quite possible that an analysis of the Questioner's "willingness to bet" in a game of Twenty

Questions with lies, links the two interpretations. See [92] for a first exploration. Other semantics for infinite-valued logic are given, e.g., in [223], [236] and [238].

Although many-valued logic was originally introduced in algebraic form by Łukasiewicz himself (see [139] and references therein), its definitive algebraization may be ascribed to Chang [36], [38], [40] who introduced MV-algebras and used the model theory of totally ordered abelian groups to prove the completeness of the Łukasiewicz axioms. The main ideas of Chang's completeness theorem are presented by Rosser in his early survey [219], and compared with the techniques used by Rose and Rosser for their own proof [216], "involving a frightening amount of detail" ([219, p.140]).

In 1940 and 1941, Moisil introduced another class of algebraic structures for his study of Łukasiewicz n-valued propositional calculi. These algebras are distributive lattices with a negation operation and some additional unary operations expressing modality (see [157]). While Moisil named these structures Łukasiewicz n-valued algebras, Rose proved that for $n \geq 5$, it is impossible to define Łukasiewicz's n-valued implication building on Moisil's modal operations (see [44, p.2]). In the papers [46] and [47] it is shown how Moisil's constructions can be modified so as to provide adequate algebraic counterparts of Łukasiewicz n-valued propositional calculi.

Independently of Łukasiewicz, Post in 1921 introduced his n-valued propositional calculi, see [203]. The algebraic counterparts of Post's calculi were described by Rosenbloom in 1942, [217] and named by him Post algebras of order n. In 1960 Epstein [79] investigated Post algebras from the lattice-theoretic viewpoint, (see [76] for Epstein's theory and its subsequent developments). In [44] it was shown that Post algebras of order n are obtainable from (Moisil's) Łukasiewicz algebras of order n by adding $n-2$ constant operators.

The monograph [28] presents a detailed study of Moisil algebras, Post algebras and (Moisil's) Łukasiewicz n-valued algebras.

Among the texts concerned with many-valued Łukasiewicz logics let us mention Rosser-Turquette [220], Ackermann [1], Rescher [206], Gottwald [102], Bolc-Borowik [29], and Malinowski [143]. The books by Wójcicki [244] and Hájek [112] contain chapters devoted to Łukasiewicz logic. Hähnle's monograph [108] is mainly devoted to automatic deduction in n-valued logics. Comprehensive bibliographies can be found in [157], [206] and [74] (also see the selected bibliography in [141].) The handbook chapters by Urquhart [237] and Panti [198] offer a survey of several many-valued systems. See [52] for a compact technical survey of MV-algebras and their neighbours.

10.2 Current Research Topics

We conclude with a list of active areas of current research in infinite-valued propositional logic and MV-algebras. The relevant literature is rapidly growing, and several papers are under review at the time of writing this section.

10.2.1 Product

Generalized conjunction connectives over the unit real interval (also known as T-norms) are interesting objects of study from various viewpoints. The reader may consult the monograph by Butnariu and Klement [34] and the relevant chapters in Hájek's book [112] for background. One of the merits of T-norm theory is to show that a substantial portion of the expressive power needed for applications of infinite-valued logic to control theory, probability theory, and game theory with variable coalitions would be provided by a logic incorporating a *product* connective jointly with Łukasiewicz disjunction and negation. Many people are actively pursuing this line of research, including Di Nola, Dvurečenskij, Esteva, Georgescu, Godo, Leustean, Panti, Riečan. We refer, e.g., the reader to the papers [81], [207] and to Montagna's analysis [159] of the relationships between MV-algebras "with product" and various categories of lattice-ordered rings.

A different approach is taken in [187], using tensor products—the latter perhaps being the bare minimum needed for if-then-else approximations of continuous real-valued functions. According to this approach, the Łukasiewicz calculus is as basic as groups are in algebra—and (tensor) multiplication naturally appears as *the* fulfillment of the following desideratum: Having a "conjunction connective" that distributes over the Łukasiewicz disjunction $x \oplus y$ whenever $x \odot y = 0$ (i.e., whenever $x \oplus y = x + y$). This distributivity law is a basic prerequisite to analyze and develop the expressive power needed for

- the approximation of a (continuous, real-valued, control) function by means of a disjunction of pure tensors,
- the definition, in every σ-complete MV-algebra, of such notions as "independent events", "conditional", and "product of two observables", for further applications in MV-algebraic probability theory (see below).

10.2.2 States, Observables, Probability, Partitions

Introduced in [182] and [184], states are the MV-algebraic generalization of *finitely additive* probability measures on boolean algebras. Their AF C^*-algebraic counterparts are known as "tracial states". On the other hand, countably infinitary operations are needed for the development of MV-algebraic measure theory. Accordingly, σ-complete MV-algebras and σ-additive states are systematically used in the book by Riečan and Neubrunn [211]. As shown by Riečan and his School, many important results of classical probability theory based on σ-complete boolean algebras and σ-fields of sets have interesting MV-algebraic generalizations. One more example can be found in [187].

While the theory of σ-additive MV-algebraic states is fairly well understood, random variables (alias, observables) still lack a definitive systematization in the context of MV-algebras. A number of technical problems, also involving product and infinite distributive laws are posed by the theory of continuous functions of several (joint) MV-algebraic observables. (See [208], [209] and [210] for interesting positive results). A useful tool for understanding such observables is given by the MV-algebraic generalization of the notion of boolean partition [184], [185] and [188]. An *MV-partition* in A is a multiset of linearly independent elements of A whose sum equals 1. This definition makes perfect sense, by referring to the underlying $\mathbf{Z}$-module structure of the unique lattice-ordered abelian group G with unit 1 given by $\Gamma(G,1) = A$. The joint refinability of any two MV-algebraic partitions on an MV-algebra A depends on the "ultrasimplicial property" of its associated ℓ-group G, in the sense that every finite set in G^+ is contained in the monoid generated by some *basis* $B \subseteq G^+$, i.e., a set B of positive elements that are independent in the $\mathbf{Z}$-module G. After some partial results of Elliott, Panti, Handelman and others (see [171], [186], [192], [191] and references therein), recently Marra [145] has proved that every abelian ℓ-group is ultrasimplicial.

10.2.3 Deduction

By contrast with finite-valued logic—and notwithstanding its rich algebraic structure—infinite-valued Łukasiewicz logic lacks a natural *proof theory*. Currently used proof techniques spuriously range from variants of integer programming [108] and Fourier-Motzkin elimination [240] to the calculation of level sets of McNaughton functions [189], [2]. The paper [4] is a first attempt to introduce an analytic calculus for the infinite-valued propositional logic of Łukasiewicz. Tight estimates are given for the complexity of the consequence relation, thus strengthening earlier results in [167].

Methodologies for *automated deduction* in infinite-valued Łukasiewicz logic are currently investigated by several people, including Aguzzoli, Baaz, Ciabattoni, Escalada Imaz, Fermüller, Hähnle, Lehmke, Manyiá Serres, Olivetti, Paulík, Salzer, Vojtáš, Wagner. See for instance [239], [137], [109], [80], [189], [2], [3], [240], the survey paper [111] and the handbook chapter [9]. Various notions of literal, clause, resolution are being considered, and various types of deduction procedures for "easy cases" are implemented, by analogy with the Horn case and 2-CNF case in the classical propositional calculus.

Altogether, much work is still to be done before proof theory and automated deduction in infinite-valued propositional logic reach a mature stage.

10.2.4 Further constructions

MV-algebraic *coproducts* (= MV-algebraic *free products*) are considered in [172] and by Di Nola, Lettieri in [66]. In [168] it is proved that the amalgamation property holds in the variety of MV-algebras. This result is generalized in [67].

For *automorphisms* of free MV-algebras see [71]. A moment's reflection shows that the automorphism group of the free MV-algebra with one generator consists of precisely two elements: identity and rotation of McNaughton functions around the vertical line through the point $(1/2, 0)$. By contrast, already the automorphism group of the free MV-algebra with two generators is highly nontrivial.

Order-theoretic and topological *properties of the set of ideals* of MV-algebras, as well as various kinds of representations, including sheaf representations, are studied in [22], [57], [17], [16], [72], [146], [84], [147], [148], [205], [116], [117], [118], [120], [70]. The MV-algebraic reformulation of a long-standing open problem in ℓ-group theory asks for a purely topological characterization of the possible spaces of prime ideals of MV-algebras with the natural (hull-kernel) topology. The order-theoretic characterization was given in [56].

The *relations between MV-algebras and various structures*, notably distributive lattices, ordered monoids and ℓ-groups, are investigated in [213], [214], [31], [32], [13], [173], [63], [35], [96], [196], [115], [19], [180], [78] and [59]. *Special classes* of MV-algebras are considered in [7], [21], [20], [54], [99], [100], [114], [131], [133], [134], [180], [85], [75] [18].

A *classification* of the universal classes generated by certain totally ordered MV-algebras is given in [95]. For subvarieties, quasi-varieties and related topics, see [215], [230], [248], [27], [83], [87], [94]. In this book, *finite-valued Łukasiewicz logics* have been considered only via their Lindenbaum algebras, namely Grigolia's MV_n-algebras, when dealing with the classification problem for subvarieties of MV-algebras. For more information on the algebras and proof theory of such systems see, for instance, [160], [161], [47], [48], [157], [8], [10], [11], [122], [173], [226].

As already noted, infinite-valued *"first-order"* notions have not been considered at all. A rapidly growing literature is concerned with the important problem of giving infinite-valued generalizations of the classical notions of point, set, cartesian product, union, subset, relation, function, equality, model, quantifier. For various constructs and attempts in this direction see, e.g., [41], [222], [90], [48], [110], [112] and references therein.

Bibliography

[1] ACKERMANN, R. (1967) **Introduction to Many-valued Logics**. Dover, New York.

[2] AGUZZOLI, S. (1998) The complexity of McNaughton functions of one variable. *Advances in Applied Mathematics*, **21**, p. 58-77.

[3] AGUZZOLI, S. (1998) A note on the representation of McNaughton functions by basic literals. *Soft Computing*, **2**, p. 111-115.

[4] AGUZZOLI, S., CIABATTONI, A. (2000) Finiteness in infinite-valued Łukasiewicz logic. *Journal of Logic, Language and Information.* Special issue on Logics of Uncertainty (Mundici,D., Ed.) To appear.

[5] AGUZZOLI, S., MUNDICI, D. (1994) An algorithmic desingularization of three-dimensional toric varieties. *Tohoku Mathematical Journal*, **46**, p. 557-572.

[6] AIGNER, M. (1996) Searching with Lies. *Journal of Combinatorial Theory*, Series A, **74**, p. 43-56.

[7] AMBROSIO, R., LETTIERI, A. (1993) A classification of bipartite MV-algebras. *Mathematica Japonica*, **38**, p. 111-117.

[8] ANSHAKOV, O.M., RYCHKOV, S.V. (1995) On finite-valued propositional calculi. *Notre Dame Journal of Formal Logic*, **36**, p. 606-629.

[9] BAAZ, M., FERMÜLLER, C.G., SALZER, G. (199?) Automated deduction for many-valued logics. In: **Handbook of Automated Reasoning**, (Robinson, A., Voronkov A., Eds.) Elsevier, Amsterdam. To appear.

[10] BAAZ, M., FERMÜLLER, C.G., SALZER, G., ZACH, R. (1998) Labeled calculi and finite-valued logics. *Studia Logica.* Special issue on Many-valued logics, (Mundici,D., Ed.,) **61**, p. 7-33.

[11] BAAZ, M., HÁJEK, P., KRAIJCEK, J., SVEIDA, D. (1998) Embedding logics into product logic. *Studia Logica.* Special issue on Many-valued logics, (Mundici,D., Ed.,) **61**, p. 35-47.

[12] BALBES, R., DWINGER, P. (1974) **Distributive Lattices**. Columbia: University of Missouri.

[13] BELLUCE, L.P. (1986) Semisimple algebras of infinite valued logic. *Canadian Journal of Mathematics*, **38**, p. 1356-1379.

[14] BELLUCE, L.P. (1992) Semisimple and complete MV-algebras. *Algebra Universalis*, **29**, p. 1-9.

[15] BELLUCE, L.P. (1995) α-complete MV-algebras. In: **Non-classical Logics and their Applications to Fuzzy Subsets**, (Höhle, U., Klement, E.P., Eds.) Kluwer Academic Publishers, Dordrecht, p. 7-21.

[16] BELLUCE, L.P., DI NOLA, A. (1996) Yosida type representation for perfect MV-algebras. *Mathematical Logic Quarterly*, **42**, p. 551-563.

[17] BELLUCE, L.P., SESSA, S. (1997) Orthogonal decompositions of MV-spaces. *Mathware and Soft Computing*, **4**, p. 5-22.

[18] BELLUCE, L.P., DI NOLA, A., GEORGESCU, G. (1999) Perfect MV-algebras and ℓ-rings. *Journal of Applied Nonclassical Logics*, Special issue on many-valued logic, (Carnielli, W.A., Ed.,) **9**. To appear.

[19] BELLUCE, L.P., DI NOLA, A., LETTIERI, A. (1993) Subalgebras, direct products and associated lattices of MV-algebras. *Glasgow Math. Journal*, **34**, p. 301-307.

[20] BELLUCE, L.P., DI NOLA, A., LETTIERI, A. (1993) Local MV-algebras. *Rendiconti Circolo Matematico di Palermo*, **42**, p. 347-361.

[21] BELLUCE, L.P., DI NOLA, A., SESSA, S. (1991) Triangular norms, MV-algebras and bold fuzzy set theory. *Mathematica Japonica*, **36**, p. 481-487.

[22] BELLUCE, L.P., DI NOLA, A., SESSA, S. (1994) The prime spectrum of an MV-algebra. *Mathematical Logic Quarterly*, **40**, p. 331-346.

[23] BERLEKAMP, E.R. (1968) Block coding for the binary symmetric channel with noiseless, delayless feedback. In: **Error-correcting Codes.** (Mann, H. B., Ed.) Wiley, New York, p. 330-335.

[24] BIGARD, A., KEIMEL, K., WOLFENSTEIN, S. (1971) **Groupes et Anneaux Rèticulés.** Springer-Verlag, Berlin, *Lecture Notes in Mathematics*, v. 608.

[25] BIRKHOFF, G. (1967) **Lattice Theory.** 3rd ed., American Mathematical Society, Providence.

[26] BIRKHOFF, G., VON NEUMANN, J. (1936) The Logic of Quantum Mechanics. *Annals of Mathematics*, **37**, p. 823-843.

[27] BLOK, W.J., FERREIRIM, I.M.A. (1993) Hoops and their implicational reducts. In: **Algebraic Methods in Logic and in Computer Science**, Institute of Mathematics. Polish Academy of Sciences, Warsaw, p. 219-230. (*Banach Center Publications*, v. 28)

[28] BOICESCU, V., FILIPOIU, A., GEORGESCU, G., RUDEANU, S. (1991) **Łukasiewicz-Moisil Algebras.** North-Holland, Amsterdam. (*Annals of Discrete Mathematics*, v. 49)

[29] BOLC, L., BOROWIK, P. (1992) **Many-valued Logics 1: Theoretical Foundations.** Springer-Verlag, Berlin.

[30] BORKOWSKI, L. (Ed.) (1970) **Selected Works of J. Łukasiewicz.** North-Holland, Amsterdam.

[31] BOSBACH, B. (1981) Concerning bricks. *Acta Mathematica Academiae Scientiarum Hungaricae*, **38**, p. 89-104.

[32] BOSBACH, B. (1982) Concerning cone algebras. *Algebra Universalis*, **15**, p. 58-66.

[33] BUFF, H.W. (1985) Decidable and undecidable MV-algebras. *Algebra Universalis*, **21**, p. 234-249.

[34] BUTNARIU, D., KLEMENT, E.P. (1995) **Triangular norm-based measures and games with fuzzy coalitions**. Kluwer, Dordrecht.

[35] CASARI, E. (1989) Comparative logics and abelian ℓ-groups. In: **Logic Colloquium 1988**, Padova. (Ferro, R. , Bonotto, C., Valentini,S., Zanardo, A, Eds.), North-Holland, Amsterdam, p. 161-190. (*Studies in Logic and the Foundations of Mathematics*)

[36] CHANG, C.C. (1958) Algebraic analysis of many-valued logics. *Transactions of the American Mathematical Society*, **88**, p. 467-490.

[37] CHANG, C.C. (1958a) Proof of an axiom of Łukasiewicz. *Transactions of the American Mathematical Society*, **87**, p. 55-56.

[38] CHANG, C.C. (1959) A new proof of the completeness of the Łukasiewicz axioms. *Transactions of the American Mathematical Society*, **93**, p. 74-90.

[39] CHANG, C.C. (1963) Logic with positive and negative truth values. *Acta Philosophica Fennica*, **16**, p. 19-39.

[40] CHANG, C.C. (1998) The writing of the MV-algebras. *Studia Logica*, special issue on Many-valued logics, (Mundici,D., Ed.,) **61**, p. 3-6.

[41] CHANG, C.C., KEISLER, H.J. (1966) **Continuous Model Theory**. Princeton University. (*Annals of Mathematics Studies*)

[42] CHANG, C.C., KEISLER, H.J. (1973) **Model Theory**. North-Holland, Amsterdam.

[43] CICALESE, F., MUNDICI, D. (199?) Optimal binary search with two unreliable tests and minimum adaptiveness, In: **Proceeding of the European Symposium on Algorithms, ESA'99**. *Lecture Notes in Computer Science*. To appear.

[44] CIGNOLI, R. (1970) **Moisil Algebras**. Bahía Blanca: Universidad Nacional del Sur. *Notas de Lógica Matemática*, v. 27.

[45] CIGNOLI, R. (1971) Stone filters and ideals in distributive lattices. *Bulletin Mathématique de la Societé des Sciences Mathématiques et Physiques de la République Populaire Roumaine*, **15**, p. 131-137.

[46] CIGNOLI, R. (1980) Some algebraic aspects of many-valued logics. In: **Third Brazilian Conference on Mathematical Logic**, Recife, 1979. (Arruda,A. I., da Costa, N. C. A., Sette, A. M., Eds.,) Sociedade Brasileira de Lógica, São Paulo, p. 49-69.

[47] CIGNOLI, R. (1982) Proper n-valued Łukasiewicz algebras as S-algebras of Łukasiewicz n-valued propositional calculi. *Studia Logica*, **41**, p. 3-16.

[48] CIGNOLI, R. (1984) An algebraic approach to elementary theories based on n-valued Łukasiewicz logics. *Zeitschrift für math. Logik und Grundlagen der Mathematik*, **30**, p. 87-96.

[49] CIGNOLI, R. (1991) Complete and atomic algebras of the infinite-valued Łukasiewicz logic. *Studia Logica*, **50**, p. 375-384.

[50] CIGNOLI, R. (1993) Free lattice-ordered abelian groups and varieties of MV-algebras. In: **IX Latin American Symposium on Mathematical Logic**, Bahía Blanca, p. 113-118. (*Notas de Lógica Matemática*, v. 38, part I)

[51] CIGNOLI, R., MUNDICI, D. (1997) An elementary proof of Chang's completeness theorem for the infinite-valued calculus of Łukasiewicz. *Studia Logica*, **58**, p. 79-97.

[52] CIGNOLI, R., MUNDICI, D. (1998) An invitation to Chang's MV-algebras. In: **Advances in Algebra and Model-Theory**, (M.Droste and R.Göbel, Eds.,) Gordon and Breach Publishing Group, Readings UK, p. 171-197.

[53] CIGNOLI, R., MUNDICI, D. (1998a) An elementary presentation of the equivalence between MV-algebras and ℓ-groups with strong unit. *Studia Logica*, special issue on Many-valued logics, (Mundici,D., Ed.,) **61**, p. 49-64.

[54] CIGNOLI, R., TORRENS, A. (1995) Retractive MV-algebras. *Mathware and Soft Computing*, **2**, p. 157-165.

[55] CIGNOLI, R., TORRENS, A. (1996) Boolean products of MV-algebras: hypernormal MV-algebras. *Journal of Math. Analysis and Applications*, **199**, p. 637-653.

[56] Cignoli, R., Torrens, A. (1996a) The poset of prime ℓ-ideals of an abelian ℓ-group with strong unit. *Journal of Algebra*, **184**, p. 604-612.

[57] Cignoli, R., Di Nola, A., Lettieri, A. (1991) Priestley duality and quotient lattices of many-valued algebras. *Rendiconti Circolo Matematico di Palermo*, Serie II, **40**, p. 371-384.

[58] Cignoli, R. D'Ottaviano I.M.L., Mundici, D. (1995) **Algebras of Łukasiewicz Logics**, (in Portuguese). Second Edition. Editions CLE, State University of Campinas, Campinas, S.P., Brazil. (*Coleção CLE/UNICAMP*, v. 12)

[59] Cignoli, R., Elliott, G.A., Mundici, D. (1993) Reconstructing C^*-algebras from their Murray von Neumann orders. *Advances in Mathematics*, **101**, p. 166-179.

[60] Cornish, W.H. (1980) Lattice-ordered groups and BCK-algebras. *Mathematica Japonica*, **25**, p. 471-476.

[61] Cornish, W.H. (1982) On Iséki's BCK-algebras. M. Dekker, New York, p. 101-122. (*Lecture Notes in Pure and Applied Mathematics*, v. 74)

[62] Czyzowicz, J., Mundici, D., Pelc, A. (1989) Ulam's searching game with lies. *Journal of Combinatorial Theory*, Series A, **52**, p. 62-76.

[63] Di Nola, A. (1991) Representation and reticulation by quotients of MV-algebras. *Ricerche di Matematica*, (Naples), **40**, p. 291-297.

[64] Di Nola, A. (1993) MV-algebras in the treatment of uncertainty. In: **Fuzzy Logic**, Proceedings of the International Congress IFSA, Bruxelles 1991. (Löwen P., Roubens E., Eds.,) Kluwer, Dordrecht, p. 123-131.

[65] Di Nola, A., Lettieri, A. (1994) Perfect MV-algebras are categorically equivalent to abelian ℓ-groups. *Studia Logica*, **53**, p. 417-432.

[66] Di Nola, A., Lettieri, A. (1996) Coproduct MV-algebras, nonstandard reals, and Riesz spaces. *Journal of Algebra*, **185**, p. 605-620.

[67] Di Nola, A., Lettieri, A. (199?) One chain generated varieties of MV-algebras. *Journal of Algebra*. To appear.

[68] Di Nola, A., Lettieri, A. (199?a) Equational characterization of all varieties of MV-algebras. *Journal of Algebra*. To appear.

[69] Di Nola, A., Sessa, S. (1995) On MV-algebras of continuous functions. In: **Non-classical Logics and their Applications to Fuzzy Subsets**, (Höhle, U., Klement, E. P., Eds.,) Kluwer Academic Publishers, Dordrecht, p. 23-32.

[70] Di Nola, A., Georgescu, G., Sessa, S. (199?) Closed ideals of MV-algebras. In: **Proceedings XIth Brazilian Logic Meeting**, Special issue of *Contemporary Mathematics*, American Math. Society. To appear.

[71] Di Nola, A., Grigolia, R., Panti, G. (1998) Finitely generated free MV-algebras, and their automorphism groups. *Studia Logica*, special issue on Many-valued logics, (Mundici,D., Ed.,) **61**, p. 65-78.

[72] Di Nola, A., Liguori, F., Sessa, S. (1993) Using maximal ideals in the classification of MV-algebras. *Portugalia Mathematica*, **50**, p. 87-102.

[73] D'Ottaviano, I.M.L. (1992) A Lógica Clássica e o Surgimento das Lógicas não-clássicas. In: **Século XIX: O Nascimento da Ciência contemporânea**. (Évora, F.R.R., Ed.,) University of Campinas, Campinas, S.P., Brazil, p. 65-94. (*Coleção CLE/UNICAMP*, v. 11)

[74] Dunn, J. M., Epstein, G., (Eds.) (1977) **Modern Uses of multiple valued Logics**. D. Reidel, Dordrecht.

[75] Dvurečenskij, A., Riečan, B. (199?) Weakly divisible MV-algebras and product. *Journal of Math. Analysis and Applications*. To appear.

[76] Dwinger, P. (1977) A survey of the theory of Post algebras. In: Dunn, J.M., Epstein, E., (Eds.) **Modern Uses of multiple valued Logics**. D. Reidel, Dordrecht, p. 53-75.

[77] Elliott, G.A. (1976) On the classification of inductive limits of sequences of semisimple finite-dimensional algebras. *Journal of Algebra*, **38**, p. 29-44.

[78] Elliott, G.A., Mundici, D. (1993) A characterization of lattice-ordered abelian groups. *Mathematische Zeitschrift*, **213**, p. 179-185.

[79] EPSTEIN, G. (1960) The lattice theory of Post algebras. *Transactions of the American Mathematical Society*, **95**, p. 300-317.

[80] ESCALADA IMAZ, G., MANYIÁ SERRES, F. (1994) The satisfiability problem for multiple-valued Horn formulae. In: **Proceedings of the International Symposium on Multiple-Valued Logics, ISMVL '94.** Boston, MA, USA, IEEE Press, Los Alamitos, p. 250-256.

[81] ESTEVA, F., GODO, L., MONTAGNA, F. (199?) The LP and LP1/2 logics: two complete fuzzy systems joining Łukasiewicz and product logic. *Archive for Math. Logic.* To appear.

[82] EWALD, G. (1996) **Combinatorial convexity and algebraic geometry.** Springer-Verlag, Berlin, Heidelberg, New York (*Graduate Texts in Mathematics, v. 168*)

[83] FERREIRIM, I.M.A. (1992) *On varieties and quasivarieties of hoops and their reducts.* University of Illinois, Chicago. (Ph D Thesis)

[84] FILIPOIU, A., GEORGESCU, G. (1995) Compact and Pierce representations of MV-algebras. *Romanian Journal Pure and Applied Math.*, **40**, p. 599-618.

[85] FILIPOIU, A., GEORGESCU, G., LETTIERI, A. (1997) Maximal MV-algebras. *Mathware and Soft Computing*, **4**, p. 53-62.

[86] FONT, J.M., RODRÍGUEZ, A.J., TORRENS, A. (1984) Wajsberg algebras. *Stochastica*, **8**, p. 5-31.

[87] GAITAN, H. (1991) Quasivarieties of Wajsberg algebras. *Journal of Non-classical Logic*, **8**, p. 79-101.

[88] GAREY, M.R., JOHNSON, D.S. (1979) **Computers and Intractability: a guide to the theory of NP-completeness.** W.H. Freeman and Company, San Francisco.

[89] GEORGESCU, G., LEUSTEAN, I. (1998) Convergence in perfect MV-algebras. *Journal of Math. Analysis and Applications*, **228**, p. 96-111.

[90] GEORGESCU, G., IORGULESCU, A., LEUSTEAN, I. (1998) Monadic and closure MV-algebras. *Multi-valued Logic*, **3**, p. 235-257.

[91] GEORGESCU, G., LIGUORI, F., MARTINI, G. (1997) Convergence in MV-algebras. *Mathware and Soft Computing,* **4**, p. 41-52.

[92] GERLA B. (1999) A probabilistic approach to Ulam games. *Theoretical Computer Science.* To appear.

[93] GILES, R. (1977) A non-classical logic for physics. In: WÓJCICKI, R., MALINOWSKI, G., (Eds.) **Selected Papers on Łukasiewicz Sentential Calculi**. Polish Academy of Sciences, Wrocław, p. 13-51.

[94] GISPERT, J., TORRENS, A. (1998) Quasivarieties generated by simple MV-algebras. *Studia Logica*, special issue on Many-valued logics, (Mundici,D., Ed.,) **61**, p. 79-99.

[95] GISPERT, J., MUNDICI, D. TORRENS, A. (1999) Ultraproducts of **Z** with an application to many-valued logics. *Journal of Algebra.* To appear.

[96] GIUNTINI, R. (1996) Quantum MV algebras. *Studia Logica*, **56**, p. 393-417.

[97] GIVANT, S.R., MCKENZIE, R.N. (Eds.) (1986) **Collected Papers of Alfred Tarski**, 4 volumes. Birkhäuser, Basel.

[98] GLASS, A.M.W., HOLLAND, W.C., (Eds.) (1989) **Lattice Ordered Groups**. Kluwer, Dordrecht.

[99] GLUSCHANKOF, D. (1989) *Objetos inyectivos en álgebra de la lógica.* Universidad de Buenos Aires. (Ph.D. Thesis)

[100] GLUSCHANKOF, D. (1992) Prime deductive systems and injective objects in the algebras of Łukasiewicz infinite-valued calculi. *Algebra Universalis*, **29**, p. 354-377.

[101] GOODEARL, K. (1982) **Notes on Real and Complex C^*-algebras**. Birkhäuser, Boston. (*Shiva Mathematics Series*, v. 5)

[102] GOTTWALD, S. (1989) **Mehrwertige Logik**. Akademie-Verlag, Berlin. Expanded version in English, in preparation.

[103] GRÄTZER, G. (1971) **Lattice Theory, First Concepts and Distributive Lattices**. Freeman and Company, San Francisco.

[104] GRÄTZER, G. (1979) **Universal Algebra**. 2nd ed. Springer-Verlag, New York.

[105] GRIGOLIA, R.S. (1973) An algebraic analysis of n-valued systems of Łukasiewicz-Tarski. *Proceedings of the University of Tbilisi*, A **6-7**, p. 121-132. (in Russian)

[106] GRIGOLIA, R.S. (1977) Algebraic analysis of Łukasiewicz-Tarski's n-valued logical systems. In: WÓJCICKI, R., MALINOWSKI, G., (Eds.) **Selected Papers on Łukasiewicz Sentential Calculi**. Ossolineum, Wrocław, p. 81-92.

[107] GUZICKI, W. (1990) Ulam's searching game with two lies. *Journal of Combinatorial Theory*, Series A, **54**, p. 1-19.

[108] HÄHNLE, R. (1993) **Automated Deduction in multiple-valued Logics**. Clarendon Press, Oxford.

[109] HÄHNLE, R. (1996) Exploiting data dependencies in many-valued logics. *Journal of Applied Non-Classical Logics*, **6**, p. 49-69.

[110] HÄHNLE, R. (1998) Commodious axiomatization of quantifiers in multiple-valued logic. *Studia Logica*, special issue on Many-valued logics, (Mundici,D., Ed.,) **61**, p. 101-121.

[111] HÄHNLE, R., ESCALADA-IMAZ, G. (1997) Deduction in many-valued logics: a Survey. *Mathware and Soft Computing*, **4**, p. 69-97.

[112] HÁJEK, P. (1998) **Metamathematics of fuzzy logic**. Kluwer, Dordrecht. (*Trends in Logic, Studia Logica Library*)

[113] HAY, L.S. (1963) Axiomatization of the infinite-valued predicate calculus. *The Journal of Symbolic Logic*, **28**, p. 77-86.

[114] HÖHLE, U. (1995) Enriched MV-algebras. *Mathware and Soft Computing*, **2**, p. 167-180.

[115] HÖHLE, U. (1995a) Commutative, residuated, ℓ-monoids. In: **Non-classical Logics and their Applications to Fuzzy Subsets**, (Höhle, U., Klement, E. P., Eds.,) Kluwer Academic Publishers, Dordrecht, p. 53-106.

[116] Hoo, C.S. (1987) Bounded commutative BCK-algebras satisfying D.C.C. *Mathematica Japonica*, **32**, p. 217-225.

[117] Hoo, C.S. (1989) MV-algebras, ideals and semisimplicity. *Mathematica Japonica*, **34**, p. 563-583.

[118] Hoo, C.S. (1995) Maximal and essential ideals of MV-algebras. *Mathware and Soft Computing*, **2**, p. 181-196.

[119] Hoo, C.S. (1997) Topological MV-algebras. *Topology and its Applications* **81**, p. 103-121.

[120] Hoo, C.S., Ramana Murty, P.V. (1987) The ideals of a bounded commutative BCK-algebra. *Mathematica Japonica*, **32**, p. 723-733.

[121] Iséki, K., Tanaka, S. (1978) An introduction to the theory of BCK-algebras. *Mathematica Japonica*, **23**, p. 1-26.

[122] Iturrioz, L. (1977) An axiom system for three-valued Łukasiewicz propositional calculus. *Notre Dame Journal of Formal Logic*, **18**, p. 616-620.

[123] Jakubík, J. (1994) Direct product decomposition of MV-algebras. *Czechoslovak Mathematical Journal*, **44**, p. 725-739.

[124] Jakubík, J. (1995) On complete MV-algebras. *Czechoslovak Mathematical Journal*, **45**, p. 473-480.

[125] Jakubík, J. (1995) Sequential convergences on MV-algebras. *Czechoslovak Mathematical Journal*, **45**, p. 709-726.

[126] Jakubík, J. (1998) On archimedean MV-algebras. *Czechoslovak Mathematical Journal*, **48**, p. 575-582.

[127] Jakubík, J. (1998) Complete generators and maximal completions of MV-algebras. *Czechoslovak Mathematical Journal*, **48**, p. 597-608.

[128] Kawano, Y. (1994) On the structure of complete MV-algebras. *Journal of Algebra*, **163**, p. 773-776.

[129] Komori, Y. (1978) The separation theorem of the $\aleph_o$-valued Łukasiewicz propositional logic. *Reports of the Faculty of Sciences, Shizuoka University*, **12**, p. 1-5.

[130] KOMORI, Y. (1981) Super Łukasiewicz propositional logics. *Nagoya Mathematical Journal*, **84**, p. 119-133.

[131] LACAVA, F. (1979) Alcune proprietá delle Ł-algebre e delle Ł-algebre esistenzialmente chiuse. *Bollettino Unione Matematica Italiana*, A(5), **16**, p. 360-366.

[132] LACAVA, F. (1979a) Sulla struttura delle Ł-algebre. *Atti Accademia Nazionale dei Lincei: Rendiconti Classe di Scienze fisiche, matematiche e naturali*, **67**, p. 275-281.

[133] LACAVA, F. (1980) Sulla classe delle Ł-algebre esistenzialmente chiuse. *Atti Accademia Nazionale dei Lincei: Rendiconti Classe di Scienze fisiche, matematiche e naturali*, **68**, p. 319-322.

[134] LACAVA, F. (1989) Sulle Ł-algebre iniettive. *Bollettino Unione Matematica Italiana*, 3-A, **7**, p. 319-324.

[135] LACAVA, F., SAELI, D. (1976) Proprietà e model-completamento di alcune varietà di algebre di Łukasiewicz. *Atti Accademia Nazionale dei Lincei: Rendiconti Classe di Scienze fisiche, matematiche e naturali*, **60**, p. 359-367.

[136] LACAVA, F., SAELI, D. (1977) Sul model-completamento della teoria delle Ł-catene. *Bollettino Unione Matematica Italiana*, 14-A, **5**, p. 107-110.

[137] LEHMKE, S. (1995) *Resolution in many-valued logics*. University of Dortmund, Germany. (Dissertation)

[138] ŁUKASIEWICZ, J. (1920) O logice trójwarkościowej [On three-valued logic]. *Ruch Filozoficzny*, **6**, p. 170-171. Translated into English by Wojtasiewicz, O., in: BORKOWSKI 1970 and, by Hiż, H., in: MCCALL 1967.

[139] ŁUKASIEWICZ, J., TARSKI, A. (1930) Untersuchungen über den Aussagenkalkül. *Comptes Rendus des séances de la Société des Sciences et des Lettres de Varsovie*, Classe III, **23**, p. 30-50. Reprinted in: GIVANT & MCKENZIE 1986; English translation in: TARSKI 1956 and in: BORKOWSKI 1970.

[140] MAC WILLIAMS, J.F., SLOANE, N.J.A. (1986) **The Theory of Error-correcting Codes**. 5th ed. North-Holland, Amsterdam. (*Mathematical Library*)

[141] MALINOWSKI, G. (1977) Bibliography of Łukasiewicz logics. In: WÓJCICKI, R., MALINOWSKI, G., (Eds.) **Selected Papers on Łukasiewicz Sentential Calculi**. Polish Academy of Sciences, Wrocław, p. 189-199.

[142] MALINOWSKI, G. (1977a) S-algebras for n-valued sentential calculi of Łukasiewicz. The degrees of maximality of some Łukasiewicz logics. In: WÓJCICKI, R., MALINOWSKI, G. (Eds.) **Selected Papers on Łukasiewicz Sentential Calculi**. Polish Academy of Sciences, Wrocław, p. 149-159.

[143] MALINOWSKI, G. (1993) **Many-valued logics**. Oxford University Press, Clarendon Press, Oxford.

[144] MANGANI, P. (1973) Su certe algebre connesse con logiche a piú valori. *Bollettino Unione Matematica Italiana* (4), **8**, p. 68-78.

[145] MARRA, V. (199?) Every abelian ℓ-group is ultrasimplicial. *Journal of Algebra*. To appear.

[146] MARTÍNEZ, N.G. (1990) *Una dualidad topológica para estructuras algebraicas ordenadas*. Universidad de Buenos Aires. (Ph.D. Thesis)

[147] MARTÍNEZ, N.G. (1990a) The Priestley duality for Wajsberg algebras. *Studia Logica*, **49**, p. 31-46.

[148] MARTÍNEZ, N.G., PRIESTLEY, H.A.P. (1995) Uniqueness of MV-algebra implication and De Morgan negation. *Mathware and Soft Computing*, **2**, p. 229-245.

[149] MCCALL, S. (Ed.) (1967) **Polish Logic 1920-1939**. Clarendon Press, Oxford.

[150] MCKENZIE, R.N., MCNULTY, G.F., TAYLOR, W. F. (1987) **Algebras, Lattices, Varieties**. Wadsworth & Brooks: Cole Publ., Monterrey, v. 1.

[151] MCLANE, S. (1971) **Categories for the Working Mathematician**. Springer-Verlag, Berlin.

[152] McNaughton, R. (1951) A theorem about infinite-valued sentential logic. *The Journal of Symbolic Logic*, **16**, p. 1-13.

[153] Menu, J., Pavelka, J. (1976) A note on tensor products of the unit interval. *Commentarii Math. Univ. Carolinae*, **17**, p. 71-83.

[154] Meredith, C.A. (1958) The dependence of an axiom of Łukasiewicz. *Transactions of the American Mathematical Society*, **87**, p. 54.

[155] Moisil, G. (1940) Recherches sur les logiques nonchrysippiènnes. *Annales Scientifiques de l' Université de Jassy*, **26**, p. 431-436. Reprinted in: Moisil, G. 1972, p. 195-232.

[156] Moisil, G. (1941) Notes sur les logiques nonchrysippiènnes. *Annales Scientifiques de l'Université de Jassy*, **27**, p. 86-98. Reprinted in: Moisil, G. (1972), p. 233-243.

[157] Moisil, G. (1972) **Essays sur les Logiques Nonchrysippiènnes**. Bucharest: Académie de la République Socialiste de Roumanie.

[158] Monk, D. (1976) **Mathematical Logic**. Springer, Berlin, New York.

[159] Montagna F. (2000) An algebraic approach to propositional fuzzy logic. *Journal of Logic, Language and Information*. Special issue on Logics of Uncertainty (Mundici,D., Ed.) To appear.

[160] Monteiro, A. (1964) **Sur la Définition des Algèbres de Łukasiewicz trivalentes**. Universidad Nacional del Sur, Bahía Blanca. *Notas de Lógica Matemática*, v. 21.

[161] Monteiro, A. (1967) Construction des Algèbres de Łukasiewicz trivalentes dans les Algèbres de Boole Monadiques-I. *Mathematica Japonica*, **12**, p. 1-23. (Reprinted in: *Notas de Lógica Matemática*, Bahía Blanca, v. 11, 1974)

[162] Morelli, R. (1996) The birational geometry of toric varieties. *Journal Algebraic Geometry*, **5**, p. 751 - 782. Errata: Homepage at the University of Utah (1997), p. 767-770.

[163] Mundici, D. (1986) Interpretation of AF C^*-algebras in Łukasiewicz sentential calculus. *Journal of Functional Analysis*, **65**, p. 15-63.

[164] MUNDICI, D. (1986a) MV-algebras are categorically equivalent to bounded commutative BCK-algebras. *Mathematica Japonica*, **31**, p. 889-894.

[165] MUNDICI, D. (1986b) Mapping abelian l-groups with strong unit one-one into MV-algebras. *Journal of Algebra*, **98**, p. 76-81.

[166] MUNDICI, D. (1987) The Turing complexity of AF C^*-algebras with lattice-ordered K_0. *Lecture Notes in Computer Science*, **270**, p. 256-264.

[167] MUNDICI, D. (1987a) Satisfiability in many-valued sentential logic is NP-complete. *Theoretical Computer Science*, **52**, p. 145-153.

[168] MUNDICI, D. (1987b) Bounded commutative BCK-algebras have the amalgamation property. *Mathematica Japonica*, **32**, p. 279-282.

[169] MUNDICI, D. (1987c) Every abelian l-group with two positive generators is ultrasimplicial. *Journal of Algebra*, **105**, p. 236-241.

[170] MUNDICI, D. (1988) The derivative of truth in Łukasiewicz sentential calculus. *Contemporary Mathematics*, American Mathematical Society, **69**, p. 209-227.

[171] MUNDICI, D. (1988a) Farey stellar subdivisions, ultrasimplicial groups and K_o of AF C^*-algebras. *Advances in Mathematics*, **68**, p. 23-39.

[172] MUNDICI, D. (1988b) Free products in the category of abelian ℓ-groups with strong unit. *Journal of Algebra*, **113**, p. 89-109.

[173] MUNDICI, D. (1989) The C^*-algebras of three-valued logics. In: **Logic Colloquium 1988**, Padova. (Ferro, R., Bonotto, C., Valentini, S., Zanardo, A., Eds.), North-Holland, Amsterdam, p. 61-77. (*Studies in Logic and the Foundations of Mathematics*)

[174] MUNDICI, D. (1991) Complexity of adaptive error-correcting codes. *Lecture Notes in Computer Science*, **533**, p. 300-307.

[175] MUNDICI, D. (1992) The logic of Ulam's game with lies. In: **Knowledge, Belief and Strategic Interaction**. (Bicchieri, C., Dalla Chiara, M.L., Eds.,) Cambridge University Press, p. 275-284. (*Cambridge Studies in Probability, Induction and Decision Theory*)

[176] MUNDICI, D. (1992a) Normal forms in infinite-valued logic: the case of one variable. *Lecture Notes in Computer Science*, **626**, p. 272-277.

[177] MUNDICI, D. (1992b) Turing complexity of the Behncke-Leptin C^*-algebras with a two-point dual. *Annals of Mathematics and Artificial Intelligence*, **26**, p. 287-294.

[178] MUNDICI, D. (1993) Logic of infinite quantum systems. *International Journal of Theoretical Physics*, **32**, p. 1941-1955.

[179] MUNDICI, D. (1993a) Ulam's game, Łukasiewicz logic and AF C^*-algebras. *Fundamenta Informaticae*, **18**, p. 151-161.

[180] MUNDICI, D. (1993b) K_o, relative dimension, and the C^*-algebras of Post logic. In: **IX Latin American Symposium on Mathematical Logic**, Bahía Blanca, Argentina. *Notas de Lógica Matemática*, v. 38, part I, p. 83-92.

[181] MUNDICI, D. (1994) A constructive proof of McNaughton's Theorem in infinite-valued logics. *The Journal of Symbolic Logic*, **59**, p. 596-602.

[182] MUNDICI, D. (1995) Averaging the truth value in Łukasiewicz sentential logic. *Studia Logica*, Special issue in honor of Helena Rasiowa. **55**, p. 113-127.

[183] MUNDICI, D. (1996) Łukasiewicz normal forms and toric desingularizations. In: **Proceedings of Logic Colloquium 1993**, Keele, England. (Hodges, W., et al., Eds.,) Oxford University Press, p. 401–423.

[184] MUNDICI, D. (1996a) Uncertainty measures in MV-algebras and states of AF C^*-algebras. *Notas de la Sociedad de Matematica de Chile*, Special issue in memoriam Rolando Chuaqui, **15**, p. 42-54.

[185] MUNDICI, D. (1998) Nonboolean partitions and their logic, In: **First Springer-Verlag Forum on Soft Computing**. Prague, August 1997, *Soft Computing* **2**, p. 18-22.

[186] MUNDICI, D. (1999) Classes of ultrasimplicial lattice-ordered abelian groups. *Journal of Algebra,* **213**, p. 596-603.

[187] MUNDICI, D. (1999a) Tensor products and the Loomis-Sikorski theorem for MV-algebras. *Advances in Applied Mathematics,* **22**, p. 227-248.

[188] MUNDICI, D. (1999b) Ulam game, the logic of Maxsat, and many-valued partitions. In: **Fuzzy Sets, Logics, and Reasoning about Knowledge**, (Dubois, D. , Prade, H., Klement, E.P., Eds.), Kluwer, Dordrecht, 1999. To appear.

[189] MUNDICI, D., OLIVETTI, N. (1998) Resolution and model building in the infinite-valued calculus of Łukasiewicz. *Theoretical Computer Science,* **200**, p. 335-366.

[190] MUNDICI, D., PANTI, G. (1993) Extending addition in Elliott's local semigroup. *Journal of Functional Analysis*, **171**, p. 461-472.

[191] MUNDICI, D., PANTI, G. (1999) A constructive proof that every 3-generated ℓ-group is ultrasimplicial. In: **Logic, Algebra and Computer Science** Polish Academy of Sciences, Warszawa, p. 169-178. *(Banach Center Publications, vol. 46)*

[192] MUNDICI, D., PANTI, G. (1999a) Twenty questions with many-valued answers, In: **Models, Algebras and Proofs**, Selected papers of the Xth Latin American Symposium on Mathematical Logic, held in Bogotá, in memoriam Rolando Chuaqui, (Caicedo, X., and Montenegro, C.H., Eds.,) Marcel Dekker, Inc., New York, Basel, p. 295-313.

[193] MUNDICI, D., PASQUETTO, M. (1995) A proof of the completeness of the infinite-valued calculus of Łukasiewicz with one variable. In: **Non-classical Logics and their Applications to Fuzzy Subsets**, (Höhle, U., Klement, E.P., Eds.), Kluwer Academic Publishers, Dordrecht, p. 107-123.

[194] MUNDICI, D., TROMBETTA, A. (1997) Optimal comparison strategies in Ulam's searching game with two errors. *Theoretical Computer Science*, **182**, p. 217-232.

[195] ODA, T. (1988) **Convex Bodies and Algebraic Geometry**. Springer-Verlag, Berlin, New York. (*Ergebnisse der Math. Grenzgeb.* (3), v. 15)

[196] ONO, H. (199?) Logics without contraction rule and residuated lattices, I. Manuscript.

[197] PANTI, G. (1995) A geometric proof of the completeness of the Łukasiewicz calculus. *Journal of Symbolic Logic*, **60**, p. 563-578.

[198] PANTI, G. (1998) Multi-valued Logics. In: **Quantified Representation of Uncertainty and Imprecision**, vol. 1, (Smets, P., Ed.), Kluwer, Dordrecht, p. 25-74.

[199] PANTI, G. (1999) Varieties of MV-algebras. *Journal of Applied Non-classical Logics*, Special issue on many-valued logics. (Carnielli, W.A., Ed.), **9**, p. 141-157.

[200] PATZIG, G. (1973) Aristotle, Łukasiewicz and the origins of many-valued logic. In: **Logic, Methodology and Philosophy of Science, IV**, (Suppes, P., et al., Eds.), North-Holland, Amsterdam, p. 921-929. (*Studies in Logic and the Foundations of Mathematics,* vol. 74)

[201] PAVELKA, J. (1979) On fuzzy logic, I, II, III. *Zeitschrift für math. Logik und Grundlagen der Mathematik*, **25**, p. 45-52, 119-134, 447-464.

[202] PELC, A. (1987) Solution of Ulam's problem on searching with a lie. *Journal of Combinatorial Theory*, Series A, **44**, p. 129-140.

[203] POST, E.L. (1921) Introduction to a general theory of elementary propositions. *American Journal of Mathematics*, **43**, p. 163-185.

[204] PRIEST, G., ROUTLEY, R. (1989) First Historical Introduction: a preliminary History of paraconsistent and dialectic Approaches. In: (Priest, G., Routley, R., Norman, J., Eds.,) **Paraconsistent Logic: Essays on the Inconsistent**. Philosophia-Verlag, Munich, p. 3-75.

[205] RAY, S., SESSA, S. (199?) Representation of an MV-algebra by its triangular norms. *International Journal of Approximate Reasoning.* To appear.

[206] RESCHER, N. (1969) **Many-valued Logic**. McGraw Hill, New York.

[207] RIEČAN, B. (1999) On the product MV-algebras. *Tatra Mountain Math. Publications,* **16**. To appear.

[208] RIEČAN, B. (199?) On the joint distribution of observables. *Soft Computing.* To appear.

[209] RIEČAN, B. (199?a) On the L^p space of observables on product MV-algebras. Manuscript.

[210] RIEČAN, B. (199?b) On the probability theory on MV-algebras. *Soft Computing*. To appear.

[211] RIEČAN, B., NEUBRUNN, T. (1997) **Integral, Measure, and Ordering**. Kluwer, Dordrecht.

[212] RODRÍGUEZ, A.J. (1980) *Un Estudio algebraico de los Cálculos proposicionales de Łukasiewicz*. Universidad de Barcelona. (Ph.D. Thesis)

[213] ROMANOWSKA, A., TRACZYK, T. (1980) On commutative BCK-algebras. *Mathematica Japonica*, **24**, p. 567-583.

[214] ROMANOWSKA, A., TRACZYK, T. (1982) Commutative BCK-algebras: subdirectly irreducible algebras and varieties. *Mathematica Japonica*, **27**, p. 35-48.

[215] ROSE, A. (1953) The degree of completeness of the $\aleph_0$-valued Łukasiewicz propositional calculus. *The Journal of the London Mathematical Society*, **28**, p. 176-184.

[216] ROSE, A., ROSSER, J.B. (1958) Fragments of many-valued statement calculi. *Transactions of the American Mathematical Society*, **87**, p. 1-53.

[217] ROSENBLOOM, P. C. (1942) Post algebras, I: postulates and general theory. *American Journal of Mathematics*, **64**, p. 167-188.

[218] ROSSER, J.B. (1941) On the Many-valued Logics. *American Journal of Physics*, **9**, p. 207-212.

[219] ROSSER, J.B. (1960) Axiomatization of infinite-valued logics. *Logique et Analyse*, **11-12**, p. 137-153.

[220] ROSSER, J.B., TURQUETTE, A. (1952) **Many-valued Logics**. North-Holland, Amsterdam.

[221] SAELI, D. (1975) Problemi di decisione per algebre connesse a logiche a piú valori. *Atti Accademia Nazionale dei Lincei: Rendiconti Classe di Scienze fisiche, matematiche e naturali*, **59**, p. 219-223.

[222] SCHWARTZ, D. (1980) Polyadic MV-algebras. *Zeitschrift für math. Logik und Grundlagen der Mathematik*, **26**, p. 561-564.

[223] SCOTT, D. (1974) Completeness and axiomatizability in many-valued logic. In: **Tarski Symposium**, American Mathematical Society, Providence, p. 411-435. (*Proceedings of Symposia in Pure Mathematics*, v. 25)

[224] SESSA, S., TURUNEN, E. (1998) On complete and strongly stonian MV-algebras. *Scientiae Mathematicae*, **1**, p. 23-26.

[225] SPENCER, J. (1992) Ulam's searching game with a fixed number of lies. *Theoretical Computer Science*, **95**, p. 307-321.

[226] STACHNIAK, Z. (1998) On finitely-valued inference systems. *Studia Logica*, special issue on Many-valued logics, (Mundici,D., Ed.,) **61**, p. 149-169.

[227] SURMA, S. (1977) **Logical Works of Wajsberg**. Ossolineum, Wrocław.

[228] TARSKI, A. (1956) **Logic, Semantics, Metamathematics**. Clarendon Press, Oxford. Reprinted (1983), Hackett, Indianapolis.

[229] TOKARZ, M. (1977) A method of axiomatization of Łukasiewicz logics. In: WÓJCICKI, R., MALINOWSKI, G. (Eds.) **Selected Papers on Łukasiewicz Sentential Calculi**. Polish Academy of Sciences, Wrocław, p. 113-117.

[230] TOKARZ, M. (1977a) Degrees of completeness of Łukasiewicz logics. In: WÓJCICKI, R., MALINOWSKI, G. (Eds.) **Selected Papers on Łukasiewicz Sentential Calculi**. Polish Academy of Sciences, Wrocław, p. 127-134.

[231] TORRENS, A. (1987) W-algebras which are Boolean products of members of SR[1] and CW-algebras. *Studia Logica*, **46**, p. 263-272.

[232] TORRENS, A. (1989) Boolean products of CW-algebras and pseudocomplementation. *Reports on Mathematical Logic*, **23**, p. 31-38.

[233] TRACZYK, T. (1979) On the variety of bounded commutative BCK-algebras. *Mathematica Japonica*, **24**, p. 238-292.

[234] TRACZYK, T. (1983) Free bounded commutative BCK-algebra with one free generator. *Demonstratio Mathematica*, **16**, p. 1049-1056.

[235] ULAM, S. (1976) **Adventures of a Mathematician.** Scribner's, New York.

[236] URQUHART, A. (1973) An interpretation of many-valued logic. *Zeitschr. f. math. Logik und Grundl. der Math.*, **19**, p. 111-114.

[237] URQUHART, A. (1986) Many-valued logic. In: **Handbook of Philosophical Logic, III**, (Gabbay, D., Ed.,) Reidel, Dordrecht, p. 71-116. Updated version entitled "Basic many-valued Logic". To appear.

[238] VASYUKOV, V.L. (1993) The completeness of the factor semantics for Łukasiewicz's infinite-valued logics. *Studia Logica,* **52**, p. 143-167.

[239] VOJTÁŠ, P., PAULÍK, L.(1996) Soundness and completeness of non-classical extended SLD-resolution. In: **Extensions of Logic Programming**, (Dyckhoff R., et al., Eds.), *Lecture Notes in Artificial Intelligence*, Springer, Berlin, v. 1050, p. 289-301.

[240] WAGNER, H. (1999) A new resolution calculus for the infinite-valued propositional logic of Łukasiewicz. To appear.

[241] WAJSBERG, M. (1931) Aksjomatyzacja trówartościowego rachunku zdań [Axiomatization of the three-valued propositional calculus]. *Comptes Rendus des Séances de la Societé des Sciences et des Lettres de Varsovie*, Classe 3, **24**, p. 126-148. English translation by R. Gruchman and S. McCall, in: MCCALL 1967, and in: SURMA 1977.

[242] WAJSBERG, M. (1935) Beiträge zum Metaaussagenkalkül I. *Monatshefte für Mathematik und Physik*, **42**, p. 221-242. English translation in: SURMA 1977.

[243] WÓJCICKI, R. (1973) On matrix representations of consequence operations of Łukasiewicz sentential calculi. *Zeitschrift für math. Logik und Grundlagen der Mathematik*, **19**, p. 239-247. Reprinted in: WÓJCICKI, R., MALINOWSKI, G. (Eds.) (1977) **Selected Papers on Łukasiewicz Sentential Calculi**, p. 101-111.

[244] WÓJCICKI, R. (1988) **Theory of Logical Calculi: Basic Theory of Consequence Operations.** Kluwer, Dordrecht. (*Synthese Library*, v. 199)

[245] WÓJCICKI, R., MALINOWSKI, G. (Eds.) (1977) **Selected Papers on Łukasiewicz Sentential Calculi.** Ossolineum, Wrocław.

[246] WOLEŃSKI, J. (1989) **Logic and Philosophy in the Lvov-Warsaw School.** Kluwer, Dordrecht. (*Synthese Library*, v. 198)

[247] WOLF, R.G. (1977) A survey of many-valued logics (1966-1974). In: DUNN, J. M., EPSTEIN, G. (Eds.) **Modern Uses of multiple valued Logics.** D. Reidel, Dordrecht, p. 167-323.

[248] WOZNIAKOWSKA, B. (1978) Algebraic proof of the separation theorem for the infinite-valued logic of Łukasiewicz. *Reports on Mathematical Logic*, **10**, p. 129-137.

[249] YUTANI, H. (1977) On a system of axioms of commutative BCK-algebras. *Mathematics Seminar Notes of the Kobe University*, **5**, p. 255-256.

[250] ZIEGLER, G.M. (1995) **Lectures on Polytopes.** Springer-Verlag, Berlin, Heidelberg, New York. (*Graduate Texts in Mathematics, v. 152*)

Index

TRENDS IN LOGIC

1. G. Schurz: *The Is-Ought Problem.* An Investigation in Philosophical Logic. 1997 ISBN 0-7923-4410-3
2. E. Ejerhed and S. Lindström (eds.): *Logic, Action and Cognition.* Essays in Philosophical Logic. 1997 ISBN 0-7923-4560-6
3. H. Wansing: *Displaying Modal Logic.* 1998 ISBN 0-7923-5205-X
4. P. Hájek: *Metamathematics of Fuzzy Logic.* 1998 ISBN 0-7923-5238-6
5. H.J. Ohlbach and U. Reyle (eds.): *Logic, Language and Reasoning.* Essays in Honour of Dov Gabbay. 1999 ISBN 0-7923-5687-X
6. K. Došen: *Cut Elimination in Categories.* 1999 ISBN 0-7923-5720-5
7. R.L.O. Cignoli, I.M.L. D'Ottaviano and D. Mundici (eds.): *Algebraic Foundations of many-valued Reasoning.* 2000 ISBN 0-7923-6009-5

KLUWER ACADEMIC PUBLISHERS – DORDRECHT / BOSTON / LONDON

MIX
Papier aus verantwortungsvollen Quellen
Paper from responsible sources
FSC® C105338

If you have any concerns about our products,
you can contact us on
ProductSafety@springernature.com

In case Publisher is established outside the EU,
the EU authorized representative is:
Springer Nature Customer Service Center GmbH
Europaplatz 3, 69115 Heidelberg, Germany

Printed by Libri Plureos GmbH
in Hamburg, Germany